*f*P

# PLAGUE

## THE MYSTERIOUS PAST AND TERRIFYING FUTURE OF THE WORLD'S MOST DANGEROUS DISEASE

Wendy Orent

FREE PRESS
*New York London*
*Toronto Sydney*

FREE PRESS
A Division of Simon & Schuster, Inc.
1230 Avenue of the Americas
New York, NY 10020

Copyright © 2004 by Wendy Orent
All rights reserved,
including the right of reproduction
in whole or in part in any form.

FREE PRESS and colophon are trademarks
of Simon & Schuster, Inc.

For information regarding special discounts for bulk purchases,
please contact Simon & Schuster Special Sales at
1-800-456-6798 or business@simonandschuster.com

Designed by Jeanette Olender

Manufactured in the United States of America

1 3 5 7 9 10 8 6 4 2

Library of Congress Cataloging-in-Publication Data
Orent, Wendy, date.
Plague : the mysterious past and terrifying future of the world's
most dangerous disease / Wendy Orent.
p. cm.
Includes bibliographical references and index.
1. Plague—History—Popular works. 2. Plague—Popular works. I. Title.

RC172.O746 2004
614.5'732—dc22 2004040378

ISBN 978-1-4516-9585-4

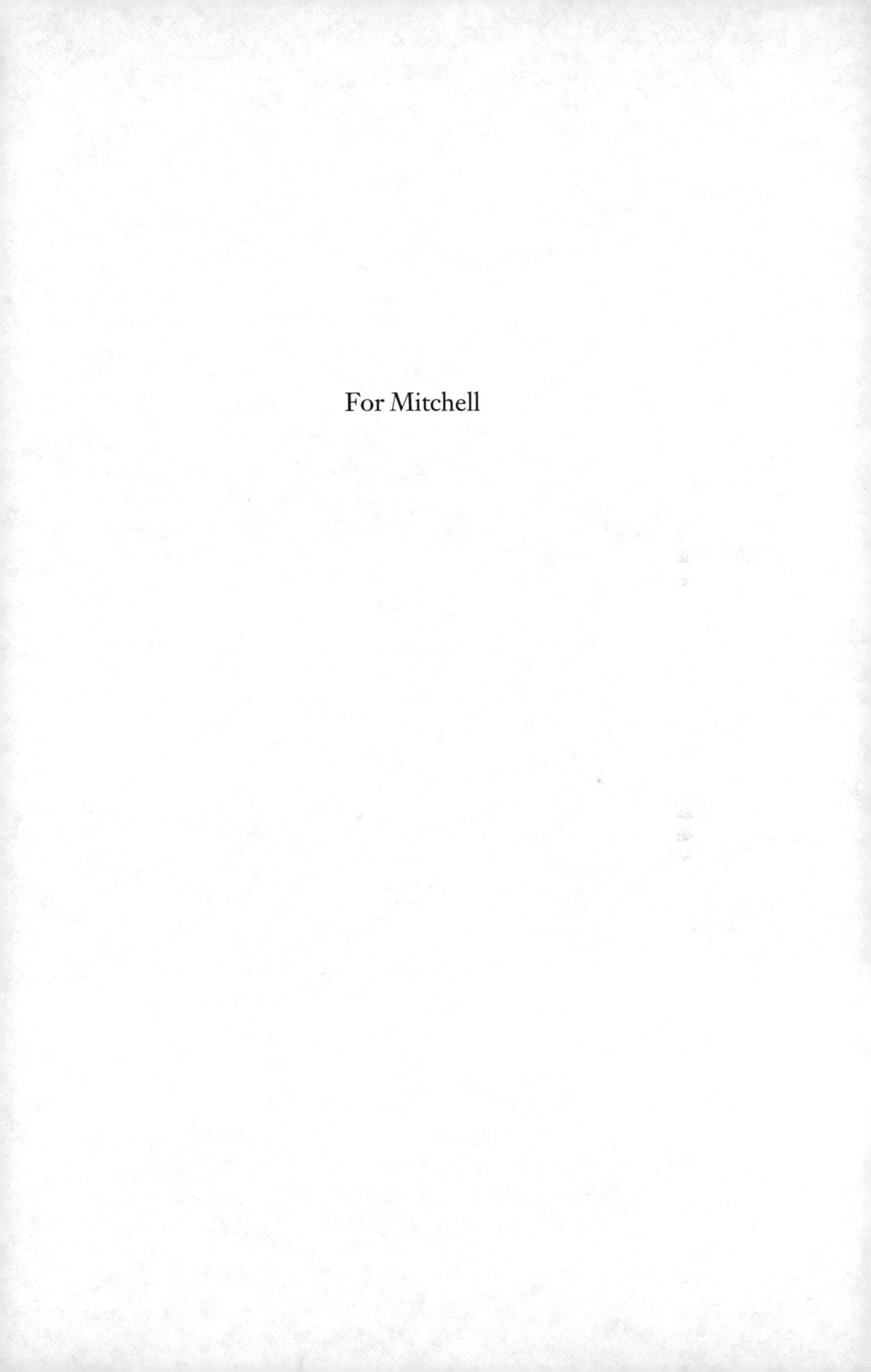

For Mitchell

# CONTENTS

Introduction, 1

I

Return to Obolensk, 7

II

The Mystery of Plague, 31

III

The Winepress of God, 61

IV

Black Death, 97

V

The Renaissance Plague, 141

VI

The Third Pandemic, 173

VII

The Enduring Threat, 209

Notes, 235

Acknowledgments, 265

Index, 269

# PLAGUE

# INTRODUCTION

One October day in 1966, a nomadic hunter in Tibet close to the Nepal border shot a tarabagan, a large, furry rodent rather like a woodchuck, a creature hunted widely in Central and East Asia for its pelt. He skinned it and brought the pelt home to his tent. Days —or perhaps only hours—later, he was burning with fever, coughing, wracked with pain. He died; his daughter, who had tended him, died soon after. So did the rest of the family, one by one; so did their neighbors, extended family members living in tents nearby.

Local authorities, hearing some disturbing reports, sent horsemen from Tibet's capital city, Lhasa, to check up on the family. After two days' hard riding, they came to the tents and looked inside. What they saw made them take their daggers and slash the ropes that bound the tents, which collapsed, becoming temporary barricades. The tents remained this way for many months, as what was inside them slowly froze in the mountain winter. Other nomadic people who lived in other tents several miles away came, took the family's abandoned sheep and cattle, and fled. When an expedition at last arrived from China, scientists found thirteen bodies. From them they recovered the frozen, and still infectious, germs of *Yersinia pestis*—plague.

Another plague story from across the world began on the 7th of February 1980. A forty-seven-year-old woman from El Dorado County, California, lost her nine-month-old pet cat to an acute infection. Three days later, the woman's own temperature shot up, but she went to work anyway, at a day-care center. The fever worsened; she developed chest pains and shortness of breath. Two days later she drove herself to the hospital. The diagnosis was pneumonia; she was treated with tetracycline, but the disease did not abate, and shortly afterward the woman died.

Not until four days after her death did medical experts realize that the woman had died of plague. Prophylactic antibiotics were belatedly rushed to the children and staff at the day-care center. Yet no one exposed to the woman fell ill.

In the difference between these two stories lies the great mystery of plague. Why is the disease sometimes sluggish, and sometimes explosive? How does it change from a lethal but virtually noncontagious disease to a slaughterer of millions?

Plague haunts our history and our literature: the catastrophe that foiled Romeo and Juliet's scheme was an outbreak of plague. Boccaccio's *Decameron* is constructed as a series of tales told during a plague outbreak by a group of wealthy young people who had fled plague-infested Florence. Monuments to the plague litter Eastern Europe. There are even plague saints: St. Roch, tended by his faithful dog, recovered from the plague and lived for several years until his martyrdom; and St. Sebastian, slain by arrows, is associated with plague because plague itself struck like an arrow of God.

We think of plague as confined safely in the past, a curse upon earlier civilizations but not ours. The Black Death burned across the medieval world like lava sweeping over Pompeii and Herculaneum, burying much of the armature of medieval civilization along with perhaps a third of Europe in its path. But epidemic plague ended in Europe only in the late seventeenth century. The days of massive depopulation, of deserted cities and heaps of dead aren't so distant. Most people are unaware that perhaps 100,000 Manchurians and 12.5 million Indians died of plague during the Third Pandemic in the twentieth century. Today plague is almost quiescent, like the banked fires of Vesuvius; but, like a curl of smoke from that mountain, the disease can still strike swiftly and subside back into its reservoir.

Plague has been studied intensively by experts all over the world, yet they still do not know why this indolent disease occasionally erupts into a virulent pandemic. No other disease has come close to plague in its death toll and its social disruption. Like Saul and David in the Bible, smallpox has slain its thousands, but plague its tens of thousands. And unlike smallpox, plague, which is adept at hiding in nature, can never be eradicated.

Scientists do not agree on where the plague germ hides between outbreaks in vertebrate animals: some say directly in soil; others say that the bacilli may be swallowed and preserved in protozoa dwelling in that soil; still others insist that the plague germ survives for months in the digestive organs of host fleas or in hibernating animals. They do agree that such hosts as gophers, marmots, ground squirrels, and rats periodically experience outbreaks of plague in particular regions where the disease is endemic. These regions are called natural plague foci, or plague reservoirs, and they are the subject of watchful attention by plague specialists all over the world.

Plague specialists consider plague a zoonosis—an animal disease that sometimes infects people. Today, except for short-term, rare outbreaks, plague no longer circulates among human hosts. Human beings contract an occasional case of it, usually from contact with an infected animal, and the story ends there, with the victim's death or recovery. Plague specialists therefore consider people to be dead-end hosts.

But the concept of dead-end host is meaningless when we think of the Black Death, where the disease, either borne from person to person by human fleas, or transmitted directly by coughing, killed a third of Europe.

At certain points in history human beings have become the principal carriers of the disease, or the chief agents of its spread. From the vantage point of the twenty-first century, we can look back and watch plague cycling in and out of human history: the Justinian Plague in early medieval times, the Black Death, the plague in Renaissance Europe, the Third Pandemic.

This book suggests that plague can evolve both greater virulence and greater transmissibility. It changes as a result of Darwinian evolution, in complex but predictable ways. A natural evolutionary process accounts for the difference between explosive plague and sluggish plague, between the Black Death and plague today. For years, disease specialists were blinkered by a single idea: that diseases, dependent as they are upon hosts for survival, grow more benign over time, as they adapt to a host species. According to this reasoning, new diseases may burst into a species with singular and lethal force, like the highly lethal hemorrhagic Ebola or Marburg viruses. Over time, diseases grow milder and milder, until, like influenza, they rarely kill at all. This approach sounds reasonable, but it is wrong. It disregards the most fundamental principle of evolution through natural selection: that the key competition involved in evolution is between individuals (or genetic lines) within a population or a species.[1] Bacterial or viral strains compete with each other, often within the ecology of a single host. Those that are best at exploiting the host's tissues win.

This blind, short-term evolutionary competition often produces ever-more lethal strains of germs. But another critical issue, from the perspective of the pathogen—a disease-causing germ—is how to get from one host into the next. If a disease is so lethal that it kills its host before it can spread, that disease will soon vanish. But there are a number of strategies for spreading available to lethal germs that do not compromise their extreme virulence. These can involve third parties—insects or even hospital attendants—to spread the infection. Or sometimes people spread lethal diseases through coughing or sneezing early in the illness, before they are immobilized. In the evolution of disease there will always be a struggle between virulence—the tendency for a germ to damage the host, even leading to host death—and transmissibility—the pathogen's ability to move on to the next host. We cannot understand the central mystery of plague, its movement from sluggishness to violent contagion, without understanding the delicate balance between virulence and transmissibility, and what requirements are necessary to set epidemic plague off again.

Long periods of quiescence, in which the plague germ hides in the soil or in the reservoirs of all-but-resistant animals, are interrupted by violent outbreaks that kill thousands of living things: this is the natural history of plague. Today, with an intense and effective global surveillance system for plague in place, and with a host of antibiotics, this natural pattern has been interrupted. Plague is still a virulent and dangerous disease, but close vigilance and effective treatment make a natural outbreak less likely than at any time in human history.

* * *

But what about unnatural plague? We now know that the disease became the Soviet military machine's chief bacteriological weapon: it is virulent, transmissible, stable, and able to accept foreign genetic information. In the years before the downfall of the Soviet Union, those scientists grew weaponized strains of plague by the ton. Indeed, if plague is ever going to strike again in a massive pandemic, it will probably be a result of deliberate human action, thanks to some of the devastating innovations developed in the Soviet Union and available to many former Soviet scientists.

This is both a natural and unnatural history of plague, because both nature and artifice have combined to make plague a terrifying threat. The story of the unnatural history of plague starts, but may not end, in the decaying laboratories of the former Soviet Union.

# I

## RETURN TO OBOLENSK

**Nothing so sweet as magic is to him,**
**Which he prefers above his chiefest bliss:**
**And this the man that in his study sits.**

CHRISTOPHER MARLOWE, *DOCTOR FAUSTUS*

*Die Seuche (The Epidemic)*. Drawing by A. Paul Weber.

©2004 ARTISTS RIGHTS SOCIETY (ARS),
NEW YORK/VG BILD-KUNST, BONN

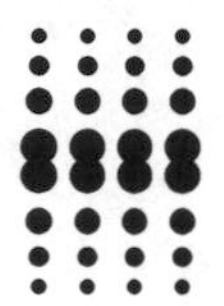

He is thin, and bowed over a little, with a pointed elfin face, and fingers slightly bent and swollen from the brucellosis he acquired from his research many years ago. But his clear, bright blue eyes are fierce with intelligence and will. When I first see him, he is sitting with his son-in-law on a bench by the customs entrance in the Moscow airport, clutching a bouquet of purple hyacinths. They have been waiting a long time—visa problems with my fifteen-year-old son have nearly kept us from entering the country. But after nearly three years of almost daily e-mails, there he is, Igor V. Domaradskij, my co-author, correspondent, and friend, whom I have never met before.

He is also perhaps the world's leading expert on *Yersinia pestis*, the germ that causes plague. It is because of plague that I have come from America to see him.

In the fall of 1999, while writing an article on Soviet bioweapons, I was tracking a mysterious and disturbing lead. I had learned that one of the principal designers of the Soviet bioweapons program was a man named Zhdanov. His name was reported in a U.S. publication as *Vladimir* Zhdanov. But the similarity of the name to Victor M. Zhdanov, a world-famous virologist and the instigator of the successful World Health Organization's smallpox eradication program, made me uneasy. Zhdanov was a genuine hero to those involved in smallpox eradication. Could he also have been a bioweaponeer?

I obtained access to a heavily scored, almost illegible copy of a typed manuscript, a Department of Defense translation of a privately published Russian-language memoir by one of the designers of the former Soviet bioweapons program, Igor V. Domaradskij. Domaradskij had been deputy director of the Interagency Science and Technology Council on Molecular Biology and Genetics, the "brain center"—as Domaradskij puts it—of the entire Soviet program. This council designed the overall aims and methods of the Soviet bioweapons system. His manuscript identified a virologist, V. Zhdanov, as the committee's chair. I obtained Domaradskij's e-mail address and started corresponding with him.

Domaradskij confirmed that the great eradicator *Victor* Zhdanov was indeed the same man who had headed the interagency council of the Soviet bioweapons program. He told me much else as well. Apparently unworried about possible repercussions from the post-Soviet Russian government, and despite some official harassment, Domaradskij was willing to share his extensive knowledge of the Soviet bioweapons program with researchers and journalists. He shared a great deal of insight and information with me. In the course of our communication we agreed to rewrite his memoir in English and attempt to find a publisher.[1] He sent me a Russian-language copy of his monograph *Chuma,* or *Plague.* This was also translated by the Department of Defense. Reading that book gave me my first glimpse of the obsession of many Russian bacteriologists with plague. They know it; they fear it; they made it—as Domaradskij admits in his memoir, if not in his monograph—into a terrifying weapon. Domaradskij's books, and our long communication, sent me on the track of plague, the world's deadliest bacterial disease, and also, to me, its most fascinating.

* * *

Domaradskij takes us to his home where he lives with his wife, the actress Svetlana Sergeevna Skortsova, still beautiful at seventy-two, their daughter Anna, and Anna's two teenaged sons. The apartment building is nondescript, even dirty, the elevator floor splashed and beaded with what looks suspiciously like dog urine. By the door in the hall an electrical wire box stands open, sprouting wires. But the flat itself, painted in soft, bright colors, coral, cream, yellow, and green, is filled with books, music, antique furniture, and paintings—of a snow-blanketed village at sunset, an old hut, an impressionistic study of three women at a well. Atop a bookcase in Domaradskij's study is what looks like a large woodchuck, paws upraised as if to ward off some coming unpleasantness. It is a stuffed marmot given to Domaradskij when he left the Anti-Plague Institute at Irkutsk for another institute at Rostov, long before his work in biological weapons design ever began. That marmot has accompanied him for forty years in his travels. Marmots are a species famous for carrying plague in Mongolia and Siberia; they are the oldest known plague reservoir in the world.

Domaradskij says that he entered the world of biological weapons because he wanted the opportunity to "do science." But this is rather a pallid way of putting it. Talking to him, reading his memoir, you sense that what propelled Domaradskij is a passion deeper than a mere liking for his work. It is a passionate hunger for knowledge, but knowledge of a certain kind. Domaradskij's work was science as creation and destruction: the power to change living things, the power over life and death. Even Faust, or Mephistopheles, could wish for nothing more.

But there is nothing devilish about this kind and cultivated man. There is no aura of evil; there is nothing striking you see in his face at all, except the intelligence and will in those blue eyes.

Domaradskij spent twenty-three years in the anti-plague system of the Soviet Union, and retains his devotion to that system, whose tragic history under Stalinist terror he has done much to elucidate. Still, he used his knowledge to turn plague, and other diseases, into antibiotic-resistant biological weapons.

There is no point asking why, or demanding an accounting, or presuming to inquire how he could fight a pathogen for much of his life, and then work to make it even deadlier and more frightening. In his memoir Domaradskij says that he never accepted the arguments for the superior humanity of one weapon over another; in other words, he might as well have been contributing to the science behind better guns, or bombs, or missiles. He also insists that he and Zhdanov thought of their work as creating "strains, not weapons." To me, he remarks that the only cannons he's ever seen are in museums. In his mind, he is truly not a "biowarrior," the American title of his memoir: he did not work on the "black arts" of bioweaponization—how to freeze-dry, stabilize, and aerosolize pathogens, or prepare them for loading into missiles.

Instead, Domaradskij solved some of the genetic mysteries of the plague germ both for the sake of the mysteries themselves, and for the secret purposes of the Soviet military. Somewhere, at some point in his life, Domaradskij crossed the bright line between studying disease and manipulating it. Now, with his repudiation of the Soviet system, his refusal to allow himself to be exploited by rogue nations (he recently refused an invitation to come to speak to scientists in Teheran "on principle"), and, most of all, his courageous willingness to tell the world what he and his colleagues have done, he has crossed it back again.

He now wants the world to know why plague is still a threat.

* * *

The day after we arrive, we go on a strange pilgrimage—to the ground zero for plague weaponeering. Nickolai, husband of Domaradskij's youngest daughter, Anastasia, is driving; Domaradskij, sitting silently beside him, smokes cigarette after cigarette, occasionally opening the window so the smoke does not fill the car. Every so often he coughs, a deep rumbling smoker's cough. My husband and I are in the back seat. We look out at Moscow, vast, ornate, and dingy; bright with the gold-leaved cupolas of churches and streets of newly opened shops, bleak with enormous bedroom communities of gray apartment buildings with row upon row of small, dirty windows. They are like massive concrete bunkers littering much of the suburban landscape leading out of the city. Along the road to Obolensk, a full two-hour drive that Domaradskij used to make several times a week ("I liked to drive then," he says), we see endless vistas of birch and pine. Intermittently there are housing developments. Some of the homes are quite large, mostly made of painted wood, and seem dropped onto empty fields in random patterns. There doesn't seem to be any landscaping, any roads, or any order to these developments; the houses perch on the ground, like a child's toys scattered across a rug. There are also older, one-room peasant cottages in bright colors—turquoise and pink and aqua—among the birch trees by the roadside; but they are lazily decaying, slowly sinking into the earth.

Domaradskij tells us how he sometimes drove this road to Obolensk with Vladimir Pasechnik, once of the Institute of Ultra-Pure Preparations in Leningrad. Pasechnik, a biochemist, was well known in the West as the first Soviet defector to reveal the existence of the mammoth, top secret Soviet bioweapons program, whose existence was long suspected by Western intelligence, but impossible to prove. Pasechnik and Domaradskij, before the former's defection, had had "good relations": Domaradskij tells us that he and Paschenik, who loved music, liked to sing Russian folk tunes as they made the two-hour-long drive together to Obolensk.

When I tell him that I have heard that Pasechnik has recently died of a sudden heart attack, Domaradskij seems surprised. But he says nothing further.

After a long while we pass through Serpukhov, the nearest town to Obolensk. It is small and bleak. Domaradskij laughs. Nickolai explains: "In the town hall, he says there is a table bearing a plaque with his picture on it. The plaque is in honor of Domaradskij's contributions to developing better vaccines to improve animal health. There was no vaccine work: this was a legend created by the masters of Obolensk to cover up their real activities, which had nothing to do with animal (or human) health." We turn off to Protvino, a small community about sixteen kilometers from Obolensk, where Domaradskij lived during his years at Obolensk; he refused to live at Obolensk itself. Protvino is beside the Oka River, in a drier and more salubrious location than the boggy, marshy Obolensk, Igor tells us. Many scientists, including physicists from a nearby laboratory, live there. We drive up to the nine-story green-painted apartment building where he had stayed in a top-floor flat. Then we drive beside the woods where, in the winter, Domaradskij went cross-country skiing. Here, he wrote in his memoir, he first began to think about the "moral implications" of his work. I remind him of that. "What did you think?" I ask. We have never discussed this before, and driving through these old haunts seems as good a time as any.

"It didn't give him insomnia," Nickolai translates. "It wasn't a spiritual crisis or depression—just thoughts."

"If you could change things, what would you do differently?" I persist.

"I would have run the lab at Obolensk the way Lev Sandakhchiev does," he replies. (Sandakhchiev is the director of Vector Laboratories—where smallpox was once weaponized.) "Sandakhchiev knows how to run a lab and keep his people. The whole bioweapons program was an adventure, and nothing much ever came out of it."

The sixteen kilometers to Obolensk consist of birch trees, and more birch trees. Sergei Popov, a defector now in the United States, who had directed Domaradskij's laboratory after the latter left the institute, once described Obolensk as surrounded by deep and gloomy forest. Listening to him, I imagined massive trees: old-growth oak, beech, pine. But here at Obolensk the sun shines through the thin, low birches, packed together like stalks of tall grass, feathered at the top with branches not yet budding. The effect is light, but somehow dingy: grey soil, grey-brown branches, narrow trees with scabby white bark. These thin grey forests, which are as deep as time, are the haunts of moose and wolves.

"You saw it at the worst time," Popov tells me later; there was neither fresh snow to cover the bleakness, nor leaves on the trees. Popov is a tall, pleasant man of fifty-two, with reddish brown hair, direct brown eyes, a ready laugh and wide smile. But his face, in repose, is lined and weary.

Above the ground outside Obolensk run pipes coated in some silver covering, which in places is slipping off. There are no checkpoints, no obvious military presence, no sense that you are approaching a formerly secret bioweapons laboratory that, now in partnership with the United States's Cooperative Threat Reduction program, still houses some of the deadliest pathogens in the world: anthrax, tularemia, Q fever, plague—many of these diseases genetically modified for vaccine and/or antibiotic resistance.

Domaradskij points out a building he identifies as Building Number One where he used to work. The building is a several-story ramshackle affair that looks like nothing so much as a slightly decrepit office building in a Third World country. This structure, which houses the most dangerous bacterial pathogens on earth, was specially designed to contain any outbreaks. Each floor is separate, Igor explains. If a breakout occurred, that floor would be completely sealed off—even the elevators wouldn't stop there. The only way for scientists to leave would be to dive into a pool of disinfectant and swim out.

"To enter Obolensk, you had to strip, shower with disinfectant, and put on your lab gown," he tells us. Afterward, you go through the whole process in reverse. Every time.

I try to take a picture, but a car drives past and they ask me to wait until it's out of sight. Domaradskij says that you never see anyone in military dress, within or outside the building, but that he's been told that presently the building is guarded by agents of both the CIA and FSB, the Russian Federal Security Service.

We pass a new housing development on the left, the same large housing we've seen intermittently littering the landscape—painted wood or stucco houses in pink and brown and red. One of the houses belongs to Major General Nickolai Urakov, Domaradskij's old enemy and the head of the institute at Obolensk. Other houses, they tell me, belong to his sons. "These homes were not allowed in Soviet times," Domaradskij says.

The large houses end—we turn right, and come to a series of crumbling apartment buildings, built in the mid-1980s, which house Obolensk's "collaborators"—the scientists who still work at the laboratory. The buildings are decaying, or perhaps were never finished: chunks of concrete are falling off, like flesh off a decaying corpse. There's a muddy road, surrounded on three sides by the birch scrubland, which circumnavigates the complex; we saw children throwing balls, mothers pushing heavy, old-fashioned prams, and young families plodding ahead on the muddy track along the roadside, apparently out for their Sunday walk; there is nowhere to go but around and around; nothing to see but the rotting buildings, one run-down shop for provisions, the unchanging scrub.

"It was so miserable to be a scientist in Russia, no money, no status," Popov says. He left Obolensk, and Russia, in 1992; it doesn't seem that much has changed.

Looking out at the complex now, Domaradskij says, "I do not feel one positive emotion, not a single good thing, coming back here again."

It is hard for the outsider to believe that this ramshackle complex, where everything seems at once unfinished and decaying, was the jewel in the crown of bacterial weapons research in Soviet times. Its location, remote and sylvan yet not too far from Moscow, was considered a benefit. To fool U.S. satellite intelligence, Obolensk was designed to look, from the air, like a sanatorium for convalescents, complete with volleyball courts.[2]

Some of the brightest stars of Soviet science were brought here.

Obolensk was a facility in the Biopreparat system, the network of biological weapons institutes supposedly under civilian control, but closely linked to the Soviet military.[3] Under control of the Fifteenth Directorate of the Soviet army, Biopreparat scientists did the military's bidding, creating disease strains possessing the characteristics—virulence, stability, antibiotic or vaccine resistance—that the military demanded.

Ken Alibek, then known as Kanatjan Alibekov, a defector well known in the United States, was also once posted to Obolensk. At the height of his power, he served as deputy director of all of Biopreparat. He defected in 1992, after the fall of the Soviet Union. Alibek earned his principal fame from the Alibekov method of weaponizing anthrax. An adept in the black arts of bioweaponry, Alibekov was extensively debriefed by shocked U.S. scientists after his defection. Before Pasechnik defected in 1989, no one in the West had any idea of the full extent of the Soviet bioweapons program. But Alibek had a great deal more to tell the West; as Domaradskij says in his memoir, he knew even more than Pasechnik. First in his classified testimony, then in newspaper accounts (beginning in 1998), and finally in his book *Biohazard*, Alibek dragged out at least something of the closed world of Biopreparat into the light. Domaradskij's memoir illuminates the beginnings of that world, which began long before Alibek ever came on the scene.

Alibek tells me that Domaradskij never had any use for him. "When I was young, Domaradskij was very famous—like a god to the younger scientists. He would look at me like I was nothing but empty space, like air." But Domaradskij and Alibek, though it was unacknowledged by either man, shared a certain outsider status among the Soviet scientific elite. Alibek was a Kazah, an ethnic group discriminated against by Russians; furthermore, his training had been at the remote laboratory at Omutninsk—"out in the sticks," as Domaradskij puts it. But in the stratified world of the supposedly classless USSR, Igor Valerianovitch Domaradskij too came of a questionable background. The descendent of Polish immigrants on one side, and the grandson of a wealthy merchant who died in Stalin's prison camps on the other, Domaradskij was a perennial outsider. Of this distinguished scientist, the widow of his former associate Victor Zhdanov once said to me, "Domaradskij! Who is he—a nothing! He came from Saratov!" Unlike her husband the famous virologist, she insisted, Domaradskij was a provincial, and not from Moscow or St. Petersburg.[4] While her comment illustrates the prejudices of the Soviet elite, she was actually wrong: Domaradskij was born in Moscow in 1925, and only consigned to the provinces with his family because his grandfather did not please the Bolshevik authorities. He grew up in the shadow of the Terror, and many members of his family died in the camps.

Enduring tyranny does not always make you a dissident, and holding outsider status doesn't always make you more sympathetic to those who share it with you. Despite a lingering distrust or even loathing of the system, Domaradskij wanted to rise within it, and to set the past behind him. By virtue of his intellect, ambition, and ferocious capacity for work, he became first a medical student and later a microbiologist. His doctoral thesis was, in his words, "sensational for the times"—and earned him a post at the prestigious Mikrob Institute, the Anti-Plague Institute in Saratov, where he set to work on the genetics of the plague microbe. He was tasked with the problem of developing a rapid method for detection of plague germs—his new method involved the use of phages, viruses that prey upon bacteria. Plague has its own phages: if the concentration of plague phages in a solution rises, then you know the solution contains plague.

In those days, he was a plague fighter, not a weaponeer. Yet the nature of scientific knowledge of plague is such that it can be used either for good or for evil. Not long after he discovered this method of detecting plague through plague phages, at the age of only thirty-one, he was offered the post of director of the Anti-Plague Institute in Irkutsk, Siberia. After some years he was offered another directorship, this time of the Anti-Plague Institute of Rostov. During his tenure at Rostov, Domaradskij first came to the attention of the Soviet Ministry of Health. Cholera broke out in the Karakalpakiya region of Uzbekistan, and a Rostov epidemic control team headed by Domaradskij was sent for to help control it. The deputy minister of health, A. I. Burnazyan, worked closely with Domaradskij and his team in the Karakalpakiya city of Nukus on containing the epidemic. In the heat and stench of their makeshift laboratories, amongst dishes spilling over with fecal samples, exposing them to the constant danger of infection, Domaradskij's team succeeded in identifying the particular strain of cholera that had broken out. His scientific success commended him to Burnazyan, as did his practical ingenuity. In order to grow colonies of cholera for identification and testing, Domaradskij needed a substance known as nutrient media. They always had it shipped from Rostov, but fresh nutrient media is perishable and it frequently spoiled in transmit. So Domaradskij devised the means to produce it himself, using fresh meat and alcohol—a process that took his team several days and attracted every chef in the area with the first-quality meat the government shipped in. Domaradskij also rigged up toilets to spray cooling jets of water over his sweltering staff, suffering from the fierce heat of Uzbekistan and the crowded, nightmarish conditions in their laboratories.

Burnazyan remembered the ingenious and dedicated young scientist, and, when Domaradskij was finally about to return home to Rostov, had him stopped at the Tashkent airport and ordered to fly directly to Moscow. There, Burnazyan stunned him with the announcement that he was to be appointed deputy minister of health himself! The appointment fell through—Domaradskij was unceremoniously dropped in favor of one General P. N. Burgasov, a favorite of Stalin's chief henchman, Lavrenty Beria. Still, he had come to the attention of the Central Committee of the Communist Party.

Meanwhile, under orders from Moscow, Domaradskij's Anti-Plague Institute at Rostov was shifting its emphasis from basic research to biological defense, an area of research the Soviets called Problem No. 5. Domaradskij developed a "dry" plague vaccine using a live strain of the disease. The Soviets frequently used live vaccine strains, which are more dangerous than the dead vaccines preferred in America, but which also frequently produce stronger immunity. Domaradskij's vaccine had the additional property of being able to withstand antibiotics, so that in an emergency it could be given concurrently with antibiotics. This was a salutary accomplishment, but there was more involved than mere prophylaxis. In order to develop an antibiotic-resistant strain of plague, Domaradskij and his team had first to understand how the plague bacterium handles new genetic information. They discovered that plasmids—rings of extra DNA that float outside the bacterium's chromosome—taken from certain intestinal bacteria such as the familiar *E. coli*, could be introduced into plague. Intestinal bacteria frequently pass such plasmids back and forth in their normal course of existence—this is an important method by which bacteria acquire new properties, including antibiotic resistance. Domaradskij's achievement in introducing antibiotic-resistant plasmids into plague allowed production of a novel vaccine, but it also demonstrated what scientists call proof of principle. If antibiotic resistance could be introduced into a living vaccine strain of plague, it could be introduced into virulent plague as well.

Eventually, Domaradskij was pulled out of Rostov, which remained on the fringes of bioweapons research, into Moscow, the center. By the early 1970s, when Domaradskij began to drift into the closed world of secret science, that world was rapidly changing. In 1969, President Richard M. Nixon unilaterally shut down the

U.S. biological weapons program. Three years later, the Biological and Toxin Weapons Convention was signed by the United States and the Soviet Union as well as seventy-seven other countries.[5] The United States indeed abandoned its program, not without the opposition of the CIA and the weapons scientists themselves. But the Soviet Union used the accord as a shield behind which they built a massive program, employing at its height perhaps thirty thousand people in dozens of secret laboratories, an empire of death that spanned the country.

Domaradskij had become well known to the powers-that-be. He was appointed by Yuri Andropov, general secretary of the Communist Party, and Leonid Brezhnev, Soviet president, as deputy chairman of the super-secret Interagency Science and Technology Council on Molecular Biology and Genetics. Zhdanov, the famous smallpox eradicator, was made chair. Domaradskij and Zhdanov were tasked with bringing the science of biological weapons into the modern age—"to catch up and leave behind" any potential enemies. This involved the swift assimilation of the genetic discoveries in the West; to that end, Zhdanov and other top scientists (although not Domaradskij) were allowed to travel to the West and to mingle with Western scientists, in order to bring back as much scientific knowledge as possible. Western scientists apparently never suspected Zhdanov's secret identity as a bioweaponeer. William Foege, the Lasker Prize–winning scientist who invented the ring vaccination strategy that eventually eliminated smallpox from nature, remembers Zhdanov as "a grandfatherly figure who had the good of the world at heart."

In 1975, Domaradskij and Zhdanov developed a plan they labeled the Five Principal Directions, which established the future course of biological weapons research. Genetic modifications of existing strains were at the heart of this program: viruses and bacteria modified for greater virulence, stability, durability in the external environment, and genetic resistance to vaccines and antibiotics. In particular, they envisioned adding short chains of proteins, or peptides, to bacteria and viruses, which would create diseases with entirely new and different symptoms, and which would make the infections more difficult to treat. They planned to divide this work up among many different institutions, some of which would concentrate on basic research, and others which were assigned to work directly on dangerous infections. Chief among these laboratories were Vector and Obolensk.

Vector Laboratories, built by convict labor in 1974, is a gigantic complex about a half-hour drive from the western Siberian city of Novosibirsk. It is dedicated to research on viruses: before 1992, it was the USSR's principal fiefdom for bioweapons research on viruses. Obolensk was the chief bacteriological weapons laboratory. It was there that, by 1980, Domaradskij found himself. He had a laboratory in Moscow, where he carried out his early work on plague virulence. He had connections to the military laboratory at Kirov, which also specialized in dangerous bacteria. But as the centripetal force of bioweapons work began to pull Domaradskij closer and closer in, he found himself working at Obolensk, at the All-Union Institute of Applied Microbiology, as it was officially designated. Appointed science director of Obolensk in 1978, he only visited his family, and his laboratory at the Institute of Protein Synthesis, back in Moscow for three days every week.[6]

At around this time, working in his own Moscow laboratory, Domaradskij made a major scientific discovery—which was never published outside the closed world of the Biopreparat system. He and his team found that plague not only could accept foreign plasmids, but had three native plasmids of its own—where most of the virulence factors are actually located. The discovery of plasmids was to revolutionize the study of plague—but Domaradskij could not tell the world about it. This loss of "his priority," as he puts it, haunts him still. The distinguished American plague researcher Robert Brubaker and his colleague D. M. Ferber first published their own discovery of plague plasmids in 1981, three years afterward. Losing the freedom to publish in the open scientific literature was a price researchers in the closed world paid for their perks and privileges: their high status, their enormous salaries, their cars, their prestige. Even in remote Protvino, Domaradskij could shop in a special grocery for the elite, where soluble coffee, fresh fruit, even caviar could be had at a time when so many in the Soviet Union stood in lines for hours for a loaf of bread.

But at Obolensk itself, life never went easily for Domaradskij. His principal project was to develop an antibiotic-resistant strain of tularemia, as he had theorized for plague.[7] Domaradskij discovered that adding antibiotic resistance disrupted the virulence of the tularemia germ. The military wing of the Soviet bioweapons program —known colloquially if rather sinisterly at Biopreparat as "The Customer"—rejected even a day's delay in the death of a test animal. The point of a weapons program is to enhance virulence, not to reduce it. So Domaradskij had to find a way around the problem, which he eventually did.

Tularemia, or rabbit fever, is a true zoonosis, a disease that comes from animals. Normally found in rabbits and rodents, it cannot transmit directly from human to human. The most virulent form, the so-called American strain, can kill 40 percent of its human victims; furthermore, it is not difficult to make the tularemia germ into a bioweapon, because it converts easily into a form that can be breathed in, and because only one or two germs produce infection. But producing antibiotic-resistant tularemia strains proved difficult; the strains rapidly lost virulence, and the more antibiotic resistance was added, the greater the problem became. Domaradskij proposed a solution: to pack two less-traumatized strains into a single weapon: each strain would only be made resistant to, say, five antibiotics, and they could be delivered together.

At the time Domaradskij came up with his binary solution, the laboratory was run by Major General Nickolai Urakov, a handsome, blunt officer of Estonian extraction. Urakov and Domaradskij clashed almost from the first. Urakov did not bother to listen; he at first dismissed this binary concept out of hand, though he was later to make extensive use of it himself.[8]

In any event, Domaradskij's hard work on the genetics of tularemia was soon to be rendered irrelevant. Tularemia was merely a sideshow, a forerunner to the main action. Though it can be deadly, it is not a contagious disease. In the early days at Obolensk, biosecurity was poorly developed; there were tularemia accidents in the laboratory, though Domaradskij's own workers were not involved. The Ministry of Health refused at first to give the Obolensk administration permission to work with more dangerous infections such as plague until certain standards were met. By various manipulations and corner-cuttings General Urakov got the laboratories up to speed, though the actual security precautions were still woefully inadequate. The Ministry of Health gave him permission for plague research, and Domaradskij's tularemia project was dropped.

* * *

Plague has always been the favorite bacterial weapon of the Soviet military—as Domaradskij puts it, after an initial bioweapons attack, "plague spreads from man to man, and further effort on the part of the military is not necessary." But Domaradskij was never allowed to work on virulent plague strains at Obolensk. Fierce, proud, and unwilling to grovel before a man he despised as a weak scientist and an arrogant, ineffective administrator, Domaradskij found himself in endless, relentless conflict with Urakov. They quarreled, first, over fundamental science. Domaradskij had always been a convinced Darwinian, a dangerous thing to be in the Soviet Union, first under Stalin, then under Khrushchev. For decades, biology in the Soviet Union had been mired in the swamp of Lysenkoism, a sort of Marxist evolutionism that spurned both natural selection and Mendelian genetics. Lysenkoists maintained that the environment could directly produce heritable changes in organisms.[9] Domaradskij and Zhdanov, who never accepted Lysenkoist dogma, agitated for the development and modernization of molecular biology and genetics in the Soviet Union. But Urakov was a Lysenkoist holdover, which did not improve his ability to run a modern microbiology institute, and which led to endless quarrels.

Second, Urakov apparently did not care to learn anything about the virulence or transmissibility or other basic biological properties of the germs studied at his institute. He wanted weapons. He was quite explicit about it, "calling a spade a spade," says Domaradskij. They fought;[10] one angry dispute followed another, and eventually, their superiors decided that the two bioweaponeeers could no longer work together. In 1987, Domaradskij left Obolensk, and eventually turned his back on Biopreparat altogether.

* * *

The Soviet Union, as an aggressive authoritarian state, was destined from the beginning to engage in bioweapons research. One afternoon in Domaradskij's study, he hands me a photograph of a commemorative plague with a series of Cyrillic names carved on it. These are the names, he tells me, of a few of the plague control scientists murdered in Stalin's purges. Domaradskij has done considerable digging to find out which scientists were murdered—information difficult to obtain. Either the younger people at the present-day Anti-Plague Institute do not know the truth of their predecessors' fate, or they prefer not to know. Professor Skorodomoff, the first director of the Irkutsk Anti-Plague Institute, was arrested and murdered. Why? Nobody knows. One of the founders of the Saratov Mikrob institute, Professor Nikanorov, was twice arrested and finally shot. "It was a terrible, horrible system," says Domaradskij's friend Lev Melnikov. "The politics of Stalin was to make people scared—people died for nothing, just to terrify other people."

Plague wasn't permitted to exist under the Stalinist regime. In the event of an outbreak, a commission, they tell me, was immediately sent to investigate it. In most cases, the commission would declare it to be "sabotage and diversion"—in other words, the direct result of some sinister plot. "People were innocent, the epidemic happened by itself, but they found some person responsible for spreading the epidemic. They pretended that some person or other was responsible for spreading plague; according to what I heard, they arrested and killed him," explains Melnikov.

According to Domaradskij, a strange sort of "liberation" for scientists occurred under Stalin's NKVD director, Lavrenty Beria, the depraved Georgian who terrorized the streets of Moscow by kidnapping, raping, and murdering young women he happened to spot on the street. Beria apparently considered it a waste of valuable resources to shoot, strangle, or poison so many scientists: he decided to put them to work instead. A system of secret prison camps, called *sharashki*, was created, where the scientists would be set tasks according to their abilities and training. Many plague control specialists were thus suborned to work for the state: but it wasn't epidemic control, presumably, they were directed to work on. It's difficult to know what exactly they did, but Domaradskij makes it clear that many of these former plague control workers applied their knowledge to the fledgling bioweapons program.

Domaradskij was never suborned in this way: he had gone into the Biopreparat system "with his eyes open," as he admits in his memoir. But he is proud to have belonged to the long tradition of plague control in the Soviet Union. Lev Melnikov tells me, for his part, that he himself never left plague control, and never worked in any capacity for the biological weapons program. The exact truth is hard to determine. According to Domaradskij's memoir, Melnikov was his Biopreparat "guardian angel," attached to the secret service. They conducted experiments together. But Melnikov explains that he was only attached to Biopreparat by assignment from the Ministry of Health, where he worked in accident prevention.

Like Domaradskij, Melnikov is a cultivated man, with an especial love of art and poetry. He grew up, like his friend, in Saratov, where they both worked at the Saratov Mikrob institute. His background touches on another reason the Soviets were born to research plague: the territorial USSR included one of the world's most virulent reservoirs of the disease. One evening at his apartment he tells me the story of a huge, almost unknown pneumonic plague outbreak that he himself witnessed in Turkmenistan, in the years 1949–1950. First, there was a huge earthquake, and then a great outbreak of plague. According to official Soviet sources, Melnikov says, only ten people died. "By my estimation, it was up into the hundreds."

"At that time there was an unusually numerous population of a rodent called *Rhombomys opimus* [the great gerbil]—it's the size of a rat, but there are some differences. This rodent forms big colonies underground. The Turkmen are nomads; they wander through the desert with their camels and their cattle, and they sleep in the sand." When plague strikes a colony of great gerbils, the fleas peculiar to the gerbils start to migrate and look for other hosts: they will bite other rodents, cattle, or human beings who might be sleeping nearby.

Melnikov, a white-haired, slightly stocky man with a wide smile showing silver teeth, was a young doctor at the time, only twenty years old. He was part of an expedition from the Saratov Anti-Plague Institute sent in to cope with the epidemic. "There was a special flight plane from Baku over the Caspian Sea, to the Afghan border. The sand was burning—you could boil an egg in it. It was 100 degrees Centigrade in the sand. The desert is deadly—it is only big dunes and saxaul trees"—bare and twisted shrubs that grow in the Central Asian desert.

"At the same time, some army units also took part in coping with the epidemic—they made a fence and guarded, blocked the infected sites. The army also had brought their specialists, but they did not dare to go to the local place where the sick were. So my friend and I—now he is dead, we were both young specialists then—we went alone. We put on special clothing—two overcoats, white overcoats. Thick masks, eyeglasses like motorcycle protective glasses, shawl-like covering, cap, rubber boots, apron, and gloves. The whole bag of instruments and everything, which we carried about one kilometer. People would not approach and bring us nearer—but we were young and we didn't care. We went to the encampment, which consisted of three large yurts."

The yurts, the large, round felt portable dwellings used by nomads in Central Asia, as they approached were completely silent. No voices, no running footsteps, no wailing of mourners or groaning of the sick. In tent after tent they saw only the silent, abandoned bodies of the dead.

In the entire encampment not one person was alive.

Melnikov believes that the epidemic began when a nomad hunter caught the infection from the great gerbils, by sleeping in the sand among the rodent colonies, where he must have been bitten by fleas. The nomad came to an encampment where he fell ill; by the time his relatives arrived for him his disease had spread to the lungs and become pneumonic. He infected them all before his death. Each one of his relatives, returning home, caused a new outbreak of pneumonic plague.

The army fell back on devices that had been used since medieval times: a strict quarantine was imposed on the Turkmen in the area. And the appurtenances of plague control—the two overcoats and the white robe over that, the masks, the gloves, the goggles—all suggest a more modern version of the plague doctor's robes in the Renaissance: the fine waxed cloth robe from head to toe, the queer long beak stuffed with spices.

"When we arrived the disease was still increasing in numbers. The precautions we took were extraordinary. The dead were burned—I am witness to this. But it was a problem—we were in the desert. There was no forest, not much wood. The military doctors said, 'Put all the bodies together with some logs, and burn everything up, together with the yurts.'" They had to bring in a big truck with oil and burn it up, to set the bodies and the yurts on fire.

With the dead stacked like logs, the black felt yurts, the deep ditch the bodies had been flung into, and the bright flames climbing into the night, it must have been a scene reminiscent of the Middle Ages. When he talks of it, you can still feel Melnikov's horror.

To commemorate the work of the plague fighters he had known, Melnikov wrote a poem that he asks me to put in English. An excerpt:

> To risk death in half-deserted aouls,
> To travel through the wilderness, the desert, where nothing grows but black saxaul,
> Who constantly keeps watch for plague, smoldering like banked embers,
> Who knows the hidden lives of innumerable rodents?
> Fighting the deadly plague, it is your obligation,
> You have met it face-to-face, and more than once.

* * *

These, then, are the two faces of plague work in Russia, the plague fighters, who endured torture, imprisonment, slaughter by bandits, and the threat of the disease itself (not a few of them died of it)—and the plague engineers, those who worked to make plague an armament. Sometimes these two faces belong to the same person.

Domaradskij feels that plague is the only bacterial disease, besides anthrax, that represents a significant bioterrorist threat.

Anthrax is dangerous because it is both lethal and durable: anthrax spores—little envelopes impervious to many environmental conditions like heat, cold, humidity, and drying—can contaminate an area, perhaps for generations. Since the mysterious anthrax attacks in late September 2001 in the United States, we have seen how easily anthrax contamination can spread throughout a building, and how difficult it is to remove the spores or decontaminate the area. But anthrax does not spread. You cannot catch it from another person, probably because anthrax is not a true pulmonary infection. It can seed itself in the lungs, but soon it moves to the lymph nodes and the chest. You can't cough it out in a contagious form: coughing produces big droplets, which can't be inhaled, and the bacteria in any event are in the vegetative—growing—state and are not infectious. Only the spores are infectious.

Plague is a very different disease from anthrax: the chief danger from plague, aside from its lethality, is its ability to spread. Unlike anthrax, it grows in the lungs; it produces infected sputum; it is coughed out and transmitted from person to person, lung to lung. Coated in some manner, it becomes more stable in the external environment: not so stable as anthrax, but stable enough. Plague germs coated with sputum, as they naturally are when they are coughed out, can last for weeks on many surfaces under certain conditions and remain infectious; plague in a frozen corpse can remain alive almost indefinitely.

Taking us on the subway back to our hotel from Domaradskij's flat, Melnikov says to me, "You must tell the American people about plague—you must scare them. Convince the American specialists that it is contagious—we know it. I have seen it with my own eyes."

# II

## THE MYSTERY OF PLAGUE

**What but design of darkness to appall?**
**If design govern in a thing so small.**

ROBERT FROST

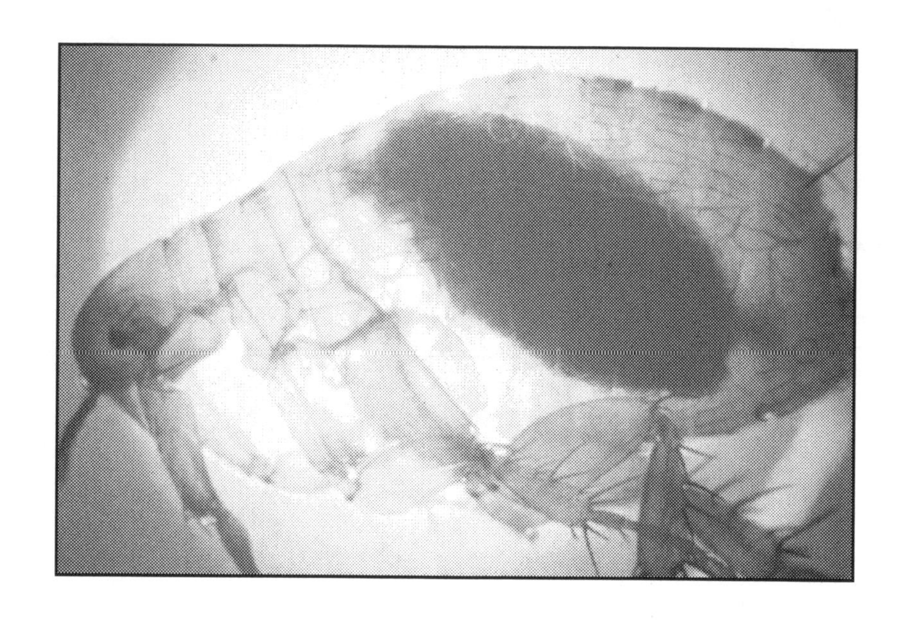

Blocked rat flea. COURTESY OF THE CENTER FOR DISEASE CONTROL

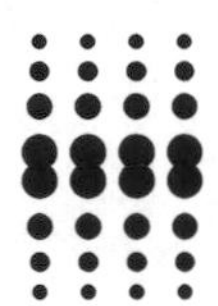

**M**elnikov's plea was an urgent one, because many experts in the West do not fear plague as much as the Russians do. American specialists have also seen plague with their own eyes. But what they have seen is very different from Melnikov's nightmare vision of burning bodies in the desert, or nomad encampments littered with the dead. Dr. Kenneth L. Gage, chief of the Plague Control Division of the Centers for Disease Control and Prevention in Fort Collins, Colorado, is frankly bemused by the attitude of scientists from the former Soviet Union toward plague. In May of 2002 he traveled to the Plague Control Center at Ulan Bator to meet with his kazakh counterparts. They showed him their plague control outfits: white suits, white masks, wrap-around head coverings, goggles. The suits resemble the garments worn by plague doctors during the Manchurian pneumonic plague epidemics of 1910 and 1920. "They actually believe plague is that contagious," he says, shaking his head. In the CDC where Gage works, plague is only considered a level two agent—which means no space suits, no biohazard bays, nothing but a lab coat, gloves, and a biosafety cabinet with a glass panel that shields the faces of technicians working with open dishes or test tubes containing the world's deadliest bacteria. They have never had a laboratory accident with plague at the CDC.

Dr. Gage, a heavyset, thoughtful man, is an aesthete, though you might not suspect it when you first meet him, or if you ride in his black pickup truck, covered in Colorado dust and littered inside with the casual detritus of family life (he has a thirteen-year-old son). But he loves birds and old books, of which he has a vast collection—some five thousand volumes. He speaks slowly, almost hesitantly at times, but his unusual diffidence does not mask his expertise—he has wide practical experience tracking plague outbreaks within and outside the United States and he knows more than anyone else in America about the ecology of plague in fleas and wild rodents.

So when Gage says that plague isn't as contagious as the Mongolian scientists think, you have to listen. Gage knows that pneumonic plague can explode in brief and terrifying outbursts. When plague swept through an Ecuadoran family in 1998, it was Gage who was called in by the World Health Organization, and Gage who stood by the seven fresh graves—"kind of a scary sight—an entire family was wiped out," he says now. A thirteen-year-old girl had slaughtered a guinea pig raised by the family for food. The girl caught pneumonic plague and died, and so, one by one, did other members of her family, including close relatives who had attended the funeral. Twelve people perished in all. Medical treatment prevented further spread, and the outbreak ended.

Still, Gage's experience is different from that of the Mongolian experts. Plague in the United States hasn't spread person to person since an outbreak in Los Angeles in 1924, when thirty-one people died of pneumonic plague. A more typical case Gage knows well happened in a small Colorado town near Fort Collins. A veterinarian contracted pneumonic plague from a sick cat brought in to her office. The veterinarian, who survived, was nursing one baby and caring for a second small child: her children must have been exposed to the disease, but she did not pass it on to them.

American plague cases, by and large, are overwhelming bubonic —they come from the bite of a plague-infected flea or from handling infected animals. Pneumonic and bubonic plague are strikingly different diseases—so different that it is amazing to realize that medieval doctors actually understood that they were two forms of one affliction. Bubonic plague is delivered, generally, by a flea, which bites a warm-blooded mammal, taking in blood and spewing forth bacilli in the next instant. Once inside the animal, *pestis* travels through the bloodstream to the lymph nodes, where it starts to replicate. Eventually the lymph nodes swell and become the huge, boggy, exquisitely painful mass we know as a bubo. Sometimes these buboes turn black and rotten, and begin to slough, revealing and destroying tissue and muscles, sometimes down to the bone. Other times, the buboes ripen and discharge large quantities of foul-smelling pus. This, in the pre-antibiotic days, represented the best hope for the patient's recovery. Even so, some 50 to 60 percent of bubonic plague victims died of the disease. Today, with streptomycin and other antibiotics, the death rate is closer to 2 percent. But the dying tissue, and sometimes the loss of fingers, nose, feet through a generalized rotting process, may still appear in bubonic plague even after it is treated: in rare instances, people may face a prolonged convalescence of months or years, or be maimed and crippled for life.

Occasionally, in the course of a malignant general infection, plague bacilli migrate toward the lungs, and the patient comes down with secondary pneumonic plague. This plague pneumonia seems to occur more frequently when people travel—a strange fact that has been noted again and again, and that probably results from the effects of stress and fatigue. It also has significance for the spread of the disease.

Generally, secondary pneumonia produces thick, ropy sputum with relatively few plague bacilli in it: this form of pneumonic plague is not terribly contagious, at least at first. But later the patient coughs up more and more blood, with more bacteria in it. If someone else breathes in a particle of that sputum—even one invisible to the eye—he will come down with primary pneumonic plague. Bubonic plague is deadly enough. But untreated pneumonic plague is almost always fatal—and sometimes, even with rapid diagnosis and intensive treatment, people still die of it.

It is hard to dispute that pneumonic plague is a contagious disease, but it is not always equally so. The woman from El Dorado County, California, mentioned in the opening pages of this book, lived in a small trailer with her sixteen-year-old daughter; she worked for two days at a day-care center; she was cared for by hospital personnel who did not know what she had. No one became sick, even though the woman was a primary pneumonic case, which is supposed to be the most contagious.

Apparently, pneumonic plague in the United States is not the same as pneumonic plague in Asia. "You are lucky people," says Lev Melnikov.

Throughout their research, the Russians have had reason to take plague seriously. In 1940, Dr. A. L. Berlin of the Saratov Mikrob institute, a plague researcher, was testing a plague vaccine strain with what's known as an aerosol challenge: he first immunized some test animals with the vaccine strain, and then sprayed live virulent plague into the air. The next day, in apparent health, he traveled from Saratov to Moscow, a train trip that took about twenty-four hours. In Moscow, he visited a barber, but then, at some point, began to feel unwell. A few days later he was dead of pneumonic plague. So were his Moscow barber and at least one of his medical attendants.

Moscow authorities realized immediately that they had an emergency on their hands. Through strict quarantine and administration of antibiotics, they ended the outbreak, but they realized how dangerous work on plague could be. From then on, research on dangerous germs was banished to remote laboratories like Vector and Obolensk.

* * *

Robert Brubaker, a Michigan State University professor who is known as the "plague guru" in the United States, is doggedly skeptical of many things, including the notion that plague differs from country to country. He is also skeptical of many Russian claims—

including that some Russian-held plague strains are notably different from the ones in American strain collections. Brubaker, like Gage, is someone whose expertise you have to reckon with. Responsible for signal contributions to our present understanding of plague biology, Brubaker was one of the first scientists in the West to discover that plague has plasmids—those extra rings of DNA outside the plague chromosome that contain some of the germ's most deadly devices.[1]

Brubaker also clarified the role of the V (for virulence) antigen, a central element in plague virulence. For a long time many plague specialists, including some of Brubaker's former students, believed that the V antigen actually served only regulatory functions for certain other factors called Yops. Brubaker insisted that the V antigen played a role in plague virulence, but found himself struggling against the current. Funding agencies rejected all of Brubaker's grant applications, and his lab was soon out of money. Brubaker had to furlough his longtime lab assistant of seventeen years in order to employ Vladimir Motin, a researcher from Russia, whom Domaradskij describes as "one of our best young scientists." Brubaker and Motin kept the lab running as they scrapped for grant money—working on other, smaller projects as they continued, unfunded, their major research on the V antigen. They showed that the V antigen contributes directly to plague virulence, and insisted that a workable vaccine must include the V antigen. Eventually, they proved to be right, with other researchers quickly confirming their findings on recombinant V antigen protection. "Brubaker's always five years ahead of everyone else," says Motin.

Brubaker is emphatic about plague's evolutionary stability and uniformity: "All plague strains are alike as two peas in a pod, they're a hell of a lot more similar to each other than two strains of salmonella and two strains of *E. coli*," he says. "It's a recently evolved pathogen and it hasn't had time to diversify." Experts like Brubaker and May Chu of the Centers for Disease Control and Prevention at

Fort Collins, Colorado, believe that the *Yersinia pestis* we see today is virtually identical to the germ that existed long ago in the days of the Black Death. They argue that it is the way we live that has changed, not the germ itself; that plague cannot take a foothold in our modern world because we are no longer the flea-ridden, rat-haunted, poverty-stricken, and malnourished people of the past. They look at the remarkable quiescence of plague in the United States for most of the past century, and think they are looking at immutable nature.

Brubaker maintains that plague virulence is indistinguishable from strain to strain—and that anyone who says differently ought to be able to prove it, in repeatable laboratory experiments designed according to the strictest scientific methods. In other words, we would have to take plague strains from around the world, sequence them, isolate their differences, and compare those exact differences by injecting them into laboratory animals and seeing what happens in the course of their infection. Only then would we know whether valid genetic differences exist—valid in the sense that they cause a different sort of disease.

Microbiologists in labs, however, do not see the plague as do observers outside the lab. The central mystery of plague has not yet been explained by any of them, despite all that may have been discovered.

* * *

The biology of the plague germ is a fearful thing, terrifying in its complexity and its power. Unlike smallpox, plague is not a specialist in human beings: it infects rats, mice, hares, camels, marmots, cats, and coyotes—over two hundred species of animals.[2] Some species have the ability to develop resistance. But perhaps because huge epidemics killing many people in the human species have only been occasional, human genetic resistance does not appear to have developed to any great extent. Plague kills, and until the advent of antibiotics, there was nothing we could do about it.

The point of medicine is to delay death long enough so that the body's own natural healing system can finally work. But against plague, particularly pneumonic or septicemic plague,[3] there is very little the body can do. The plague germ is a stealth agent, flying in under the body's radar screen; the body has no means of knowing it is under siege. There is no inflammation—*Yersinia pestis* has evolved the means to stop that protective response. The plague victim may feel quite well after the germs are already multipying throughout the body—which is why plague seems to kill so suddenly. Often there are no headaches, chills, fever, sweats, or nausea,[4] until the damage has been done, the bacteria have grown in uncounted millions, and your body has been occupied totally by the enemy.

At that point your liver, spleen, and lymph glands are no longer human organs. They are tissues of plague, plague bacteria in almost pure culture.

There is no other disease like this known to humankind. For one thing, very few diseases—and no truly human diseases at all—are so lethal. Before antibiotics, bubonic plague claimed about 60 percent of its victims, compared to the 10–50 percent lethality of smallpox, the most virulent disease that attacks only human beings. Bubonic plague is the mildest form: pneumonic and septicemic plague kill almost everyone. Even today, with antibiotics, 14 percent of plague victims in the United States still die, and still others suffer necrotic tissue, muscle destruction, the loss of toes or fingers from septicemia, and severe weakness that may last for months.

Furthermore, no other fatal epidemic disease has at least two distinct modes of transmission, producing what looks like two entirely different diseases. As pneumonic plague, infecting the lungs, it can be transmitted directly from person to person. Untreated, pneumonic plague has a fatality rate close to 100 percent. In the nonhuman world, where plague is largely a disease of rodents, it is generally transmitted through fleas. Scientists call such indirect carriers *vectors*. Vector-borne diseases allow for quite a high degree of virulence.[5] The immobilized or dying victim, perhaps too weak to swat an insect away, is bitten, and the insect flies away, bearing a toxic load of bacteria to inject into its next mammalian blood meal.

Every dog has its flea, so to speak. There are many kinds of fleas: aside from the classic plague vectors, the Oriental rat flea, and the human flea, there are lots of others. Some of these, cat fleas, dog fleas, are familiar to us from daily life. Some, less well known, represent more of a danger, particularly in the Western or mountain states: these are fleas that inhabit ground squirrels and gopher towns, either on the rodents themselves or in their burrows. (Prairie dog fleas only rarely bite people, and are an uncommon source of human plague, though plague has reduced the prairie dog population in some parts of the West by nearly 98 percent.) These rodent fleas are the principal vector of human plague in the United States.

Not all fleas transmit plague germs equally effectively. The Oriental rat flea, *Xenopsylla cheopis*, which haunts this book like a tiny destroying angel, has a peculiar talent as a carrier. It becomes blocked.[6] A ball of blood and sticky bacteria ingested from its rat host forms inside the flea's foregut, a valve connecting the flea's esophagus to its midgut. (Fleas have several parts to their stomachs.) This plague ball forms in front of the flea's stomach where it cannot be digested. It feeds more vigorously, regurgitating infected blood and flea saliva, pumping, with every fresh bite, some of the infected blood into its new host. An infected *cheopis*, under most conditions, will starve to death in a few days. But it can pass on plague germs to a new host before it dies itself.

Dr. Joe Hinnebusch of the Rocky Mountain Laboratories in Montana has studied plague in *cheopis* for years. He and his colleagues have studied a specific genetic sequence of the plague germ that produces a protein called Yersinia murine toxin (or YMT).

YMT was discovered early in the study of plague biology; when the toxin is purified and injected, it is fatal to mice. But Hinnebusch thinks that the murine toxin actually evolved for quite another purpose: to allow plague to survive in the guts of fleas.

Plague is a new disease, relatively speaking. Somewhere between 1,500 and twenty thousand years ago *Yersinia pestis* evolved from a closely related intestinal bacterium, *Yersinia pseudotuberculosis*.[7] Pseudotuberculosis is a relatively mild disorder: people who have it have sharp pains in the stomach and are sometimes operated on for appendicitis. But it's self-limiting; pseudotuberculosis is usually effectively handled by the human immune system.

Pseudotuberculosis—which can't be transmitted by a vector—can survive both in the harsh intestinal environment and in the soil. There is even a hypothesis that it can be taken up into the roots of vegetables, where an unsuspecting rabbit may dine upon it. Unlike plague, pseudotuberculosis probably benefits from a healthy host: it is perfectly happy to be shed into the soil through elimination, and acquired through ingestion.

The plague germ has lost pseudotuberculosis's tricks, but gained some lethal habits of its own. It has the ability to grow in staggering amounts at different sites throughout the body, and, unlike pseudotuberculosis, it has honed its anti-immune machinery to the point where it generally overpowers the human immune system. Through the acquiring of murine toxin and perhaps other factors, *Yersinia pestis* developed an ability to survive in the guts of fleas that altered its evolutionary course forever: no longer a benign intestinal disease, the plague germ now became a deliberate killer.[8]

The acquisition of this toxin was a first step in the evolution of plague's lethal strategy. The fleas that prey on rats, such as *cheopis*, are not like mosquitoes, diving down, biting, and flying off before an angry slap splatters them in their tracks. Rat fleas are homebodies, settled souls, who dwell much of their lives on their warm and ambulatory dinners. A rat flea may bite its host three or four times a week, biting that does not seem to much discommode the rat, unless the flea is plague-bearing and the rat susceptible.

There is an ingenious Darwinian logic at work, from the point of view of the germ. Rat fleas don't leave their host until the rat dies. Fleas only drink warm blood, and live on a warm body. Therefore, to make the rat flea leave its host and spread the disease, *Yersinia pestis* became a killer. As the body cools, the fleas jump off.[9] They prefer another rat as host, but if they can't find one they will bite other warm-blooded animals, including human beings. This, frequently, is how plague first enters the human population: if enough rats have died from plague, rat fleas of necessity turn to other warm-blooded animals living in close proximity to their original hosts.

Another important step in the evolution of the plague germ was the adaptation of the HMS gene, which *Yersinia pestis* shares with *Y. pseudotuberculosis*, to produce blockage in fleas. HMS enables germs to clump together, forming a sticky ball inside the flea's gut. The HMS gene was the discovery of one of Hinnebusch's colleagues, Robert Perry of the University of Kentucky Medical School. HMS stands for hemin storage locus: plague germs that lack this trait can't produce blockage.[10] In other words, the sticky ball of blood and germs that builds up in the gut of *cheopis* cannot form: the bacteria move directly into the flea's midgut, where they may cause a long-term infection—but they won't be passed in the flea's saliva to some other host. So blockage formation, and the hemin storage gene that causes it, are necessary if plague germs are going to use the Oriental rat flea as their means of jumping from one mammal host to another.[11]

Not all fleas can form blockages, and blockages are not necessary for fleas to be a deadly carrier. According to Ken Gage, blockage may actually be quite rare: but so many researchers get stuck on what Gage calls the "classical model" of infection through blocked rat fleas that they don't consider other alternatives. In fact, two

French plague researchers in the 1950s who worked in Kurdistan insisted that plague transmission in Kurdish villages actually occurred through unblocked fleas, in this case the so-called human flea—*Pulex irritans*. In certain parts of the world, *Pulex irritans* can infest people by the swarming hundreds.[12]

According to Gage, *Pulex irritans* is more of a nest flea than a host flea;[13] in other words, it spends more time off its host than the Oriental rat flea does. It lives in layers of clothing and on sheets; in Ecuador, when Gage went to investigate the plague deaths of an entire family, he saw clouds of *Pulex* lifting off the household bedding when it was brought into the light. Accounts from Renaissance pesthouses also describe the clouds of fleas rising off one patient and settling on another. But since *Pulex* is only likely to spread plague via what is called mechanical transmission—transporting plague germs from one host to another on its mouth parts—most experts haven't taken it seriously as a transmitter of plague. How likely is it that a flea, jumping around with a germ or two resting on its proboscis, will be able to find a new host and bite it before the germ dies?

Perhaps more likely than one might think. Though it's an inefficient carrier compared to the Oriental rat flea, *Pulex* en masse can still transmit plague.[14] Japanese biological weaponeers in World War II dropped porcelain bombs containing plague-contaminated *Pulex* over China, resulting in thousands of plague cases.[15] Unblocked fleas, also, may spread disease through mechanical transmission much more rapidly than blocked fleas do, since blockages do not have to form.[16] Gage notes that among prairie dogs plague seems to spread like wildfire, wiping out entire colonies of hundreds in a few days, too fast for blockage. Also, very few blocked prairie dog fleas have ever been found, despite field investigations mounted specifically to discover them.

Plague transmitted by unblocked fleas would be highly virulent. Not all plague cases are severe enough to produce more than a fleeting presence of germs in the blood as the bacteria move toward the spleen and liver and lymph nodes. It takes a severe case to produce enough plague germs in the bloodstream for an unblocked flea to pick it up and transmit it. The death of the first host follows—but that is immaterial to both the germ and the flea.

Merely injecting bacteria into the skin, whether by blocked or unblocked fleas, doesn't ensure that any sort of disease is going to develop. If fleas manage to pump blood and germs into the next victim, but the germs have no means of spreading and colonizing the body, the whole story ends with an undistinguished little plug of bacteria and perhaps a small sore on the skin. Another factor, plasminogen activator, is necessary for plague to move from the injection site to the regional lymph nodes, where it begins to grow in bulk, eventually forming a bubo. Once in the lymphatic system, the bacteria encounter the immune system's first line of defense: the white blood cells that normally swallow up invading bacteria or viruses and destroy them. *Yersinia pestis* has evolved an astounding array of genetic devices to combat mammalian defenses. We may have sophisticated immune systems, but plague has proven incomparably subtler and stronger in its confrontation with humanity.

First among *pestis*'s lines of attack is something called the F1 antigen, which turns on at normal mammalian temperature—and off at room temperature. The F1 antigen allows the plague germ to form a protective polymer envelope. This envelope seems to help the germ resist phagocytosis—being swallowed up by the cells that scavenge about in the body, seeking, swallowing, and destroying invaders. Lumps on the plague germ's envelope may help it to resist this swallowing.

Plague germs stuck in the foregut of a flea (at room temperature) rapidly lose this envelope. When injected into the skin of a mammal, they are initially vulnerable to immune cells. But if enough germs are injected—as they would be from the bite of a blocked rat flea—chances are the immune cells will fail to find them all at the very beginning. The germs are still swallowed by immune cells, but remain intact inside the cells, growing and multiplying in peace. Eventually they shatter the immune cell and burst out of it, spilling into the bloodstream in their multitudes. It is a Trojan horse strategy.

This particular Trojan horse, the F1 antigen, adds to the mystery of plague. Brubaker claims that the F1 antigen is trivial, since rats injected with strains lacking the F1—a few have been found in nature—die anyway. To many scientists, nothing about plague is provable, or even worth discussing, unless it's somehow demonstrable on a molecular level.

But lab researchers, microbiologists and molecular biologists, don't always consider function in nature. To the ecologist or the evolutionist, a trait that is widespread in nature must be considered to serve some function, or natural selection would have removed it from the genome. Worthless genes don't just hang around—if they aren't positively harmful, at least they are a drain on the organism's resources, and so like the State in Stalinist dogma they should wither away. And F1 hasn't done any withering.

To Sergei Balakhonov, head of the Microbiology Department of the Anti-Plague Institute of Siberia and the Far East in Irkutsk,[17] the F1 antigen is more than important—it is critical. He insists that the handful of strains found in nature that lack the antigen are actually not true *pestis* at all. "Only strains that have the F1 antigen, in my opinion, can be called *Yersinia pestis*," says Balakhonov.[18]

Current research suggests that the plague capsule's chief function is to interfere with the swallowing of germs by immune cells.[19] This interference appears to allow *pestis* to do what it does best—grow outside cells in massive, almost unthinkable bulk. Scott Bearden, a young CDC molecular biologist, believes that despite *pestis*'s ability to multiply within immune cells, it is really happier outside than inside them: it is along the network of cells lining the lymph nodes, the liver and spleen, that plague really comes into its own. It multiplies so rapidly along this cell lining that it replaces internal organs with enormous colonies of its own. This massive growth of bacteria shows how virulent a disease this is, how effectively it exploits host tissues.

F1 may also be connected with aerosol transmission—that is, from lung to lung, producing pneumonic plague, as well as via unblocked fleas. Under both circumstances, plague germs would be transmitted in a fully virulent, primed state—and many fewer germs might be necessary to start a lethal infection.[20]

Mechanical transmission via unblocked fleas remains extremely controversial. Many American scientists seem skeptical. But in Kazakhstan, plague scientists take an entirely different view. At a Yersinia meeting in Turku in September of 2002, I spoke with two Kazakh scientists from the Almaty Anti-Plague Institute, once a node in the Soviet anti-plague system, and now an independent entity under Kazakh control. In the old days, Bakyt Atshabar and Bakhtiar Suleimenov were plague researchers, though according to an American expert who knows them well, they were never involved in bioweapons work.

Suleimenov, Atshabar, and I sit at a table with a young, frosted-blond Russian translator, who gesticulates frequently with her pink manicured fingernails. Atshabar is a stocky, dark-haired man with a wide-cheekboned face that remains immobile and grave; Suleimenov, whose dark eyes are hard to read behind wide round glasses, smiles throughout, and does most of the talking. Yet of the two Atshabar seems the more approachable; he listens with a penetrating, receptive intensity.

Suleimenov insists that unblocked fleas are the principal plague vectors. Blocked fleas live for too short a time—for one to three days at the most. "Blocked fleas don't move quickly, and they don't eat much," he insists. "In nature it is very difficult to identify blocked fleas—only a few can be found in natural conditions. Unblocked fleas are healthy, they search actively for new hosts. In Kazakhstan, scientists could find blocked fleas in natural conditions, but only a few. The ratio of blocked to unblocked fleas in nature is 1:10,000.

"When unblocked fleas bite," Suleimenov continues, "they can carry four to twenty cells of the plague germ. If they replicate inside a [blocked] flea, this would be at a low temperature, and the new cells would not be in a virulent state [i.e., with the capsule expressed]. But when it's transmitted straight from one host to another, it's already in a virulent state."

These claims are heretical to most American plague specialists, and when Suleimenov is confronted with disagreement, some scientists report, he can be prickly about it. But older research confirms his claim. A 1958 study argues that blockage may be necessary to maintain plague between outbreaks, but once an outbreak begins mechanical transmission can keep it going.[21]

We can't know at this point whether the F1 antigen really evolved to aid transmission in unblocked fleas. Whatever F1's function will prove to be, it is only one of the weapons that the plague germ employs against the mammalian immune system. There are many others we know of—and no one thinks all of them have been discovered.

The myriad ways in which the plague germ overcomes its host's immune system are both fascinating and appalling. There are few things in life that seem to typify the inventive power of nature to kill more than the principal armaments of *Yersinia pestis* against the mammalian immune system. These armaments mostly fall into the category known as Yops, for *Yersinia* outer proteins.[22] Yops, whose terrifying complexities take us far beyond the scope of this book, are coded on a single plasmid, the strip of DNA that lies outside the pestis chromosome.[23] These proteins, which are found in all *Yersinia* species, belong to a class of anti-immune system weapons known as the Type Three Injector System.

This injection apparatus would look, if you could see it, like a thicket of needles springing out from the lumpy hide of *pestis* like spines from a hedgehog. Some of the Yops function like syringes—their proteins form long hollow tubes through which other cell poisons are injected into the walls of mammalian cells. It's the proximity of mammalian immune cells that triggers the needles and the toxins that pour down them.[24] A *Yersinia* cell docks next to an immune cell, sprouts a Yop needle, and injects a toxin down that needle right into its target.

Scott Bearden, the young CDC molecular biologist who trained at the prestigious University of Kentucky laboratory of Robert Perry, talks about Yops with an edgy eagerness. Immune cells normally secrete a chemical that protects them from cell death, he explains. One of the Yops prevents that chemical from working. Bearden sums up the effect: "Cells get paralyzed, they can't call for help, and then they kill themselves." When an immune cell is under attack it normally puts out chemicals called cytokines, which immediately call other immune cells to the spot to deal with the attacking invader. But a macrophage—a human immune cell, literally a "big eater"—ambushed by Yops is, effectively, gagged and murdered. This silencing of the immune system is why Robert Brubaker calls *Yersinia pestis* a "stealth agent."

One of Brubaker's important discoveries, as we have seen, is the role of still another of plague's strategic weapons, the V antigen. According to Vladimir Motin, the V antigen has several different effects. Yops have to dock right next to an immune cell to stab it with a needle. But the V antigen acts at a distance, and with a peculiar chilling efficiency: among other effects, it causes human immune cells (the helper T2 cells) to produce too much of a normal immune protein called Interleukin-10 (or IL-10).

Too much IL-10 appears to stop the production of the chemicals that produce inflammation, one of the body's chief defensive responses. Without inflammation, the body does not know it is under attack. So subtly, stealthily, the plague germ does its work, like an army stealing up into a citadel before the defenders even know the walls have been breached.

Vladimir Motin is especially fond of another of *pestis*'s tricks—one involving a certain chemical known as Lipid A. "In general, in order to develop an appropriate amount of *Yersinia pestis* in the flea gut, you have to have a reasonably high load of bugs in the blood [bacteremia]," he explains. "In general, there will be selection for high bacteremia. But this high bacteremia can cause septic shock." Lipid A is the factor that causes septic shock, which ordinarily kills the host quickly.

This is a conundrum, which *Y. pestis* has solved in its usual devious way. In order to be picked up by a flea, there have to be many "bugs in the blood." But fleas don't bite dead animals. To raise the chances of a flea bite, the animal has to be kept alive as long as possible, given all the bacteria drifting through its bloodstream. *Y. pestis* has evolved a means to have it both ways: it postpones septic shock by making Lipid A less poisonous at 37 degrees Celsius—the temperature of a living mammal—than it is in the flea at 26 degrees Celsius—room temperature. That way, despite the millions of bacteria teeming in its blood, the animal's immune system isn't "aware" that it is under attack. The animal lives longer, and the plague germ increases its chances for transmission.

What makes the plague germ so deadly, in Motin's description, is "the combination of genes, the presence or absence of certain genes, and also how they work together—how they are regulated together." He goes on to say, "So far we have not found any single thing that makes this bacteria so virulent—it's the combination of many different factors, all of which are also present in other pathogens. Somehow evolution selected for the right combinations to make this thing dedicated to kill.

"*Yersinia pestis* is the yardstick," Motin concludes.[25] It is the deadliest of all disease agents, the one by which all other germs are measured.

To understand the biology of this disease is to know its deadliness, its uniqueness, and its power. The Black Death could not have been caused by anything other than *Yersinia pestis*. No other disease can overmaster the human immune system with anything like plague's subtle devastation. No other disease can hide out in the wild, only to spring back into humanity and slaughter hundreds, or millions. No other bacterial disease poses the same threat to humanity as a biological weapon.

Fortunately, plague has never been able to permanently circulate in human populations. It is a disease that exists quietly beside us—usually unseen and unfelt. Plague may rage through a prairie dog colony in the heart of a Colorado city, leaving hundreds of the animals dead in their burrows, and never touch the jogger who runs through the town or the children who play along its edges. It is only on occasion—as history terribly demonstrates—that it makes the leap from such a reservoir. When it comes, it is devastating. Its threat is persistent, and not to be minimized. But it does not stay.

* * *

Q. Where does plague go between epizootics?

A. Into the laboratories at the Saratov Anti-plague Institute.

OLD SOVIET BIOWEAPONEER JOKE

Natural plague reservoirs—rodent populations where plague endlessly circulates—can be found in every continent of the world except Australia. A ship containing infected rats brought plague to the shores of California late in the nineteenth century: from those rats, the disease quickly spread into the ground squirrels and prairie dog populations of the Western United States and Canada. Infected rats came to Australia, too, but plague didn't find any local rodents to infect: presumably, plague doesn't like kangaroos and wombats very much. After killing a handful of people, the disease vanished from Australian shores.

But in the Old World, plague circulates very differently. Marmots, susliks (a kind of ground squirrel), and gerbils all have varying degrees of resistance: according to Domaradskij, gerbils on one side of the Volga are resistant to plague, while gerbils on the other side are highly sensitive. Domaradskij and many other researchers have suggested that the existence of resistant animals plays a major role in the circulation of plague and its maintenance in nature between epizootics—the animal equivalent of epidemics, large outbreaks that kill many animals and have the potential to spill into the human species.

Other scientists, including some of the major French specialists, insist that plague germs can remain alive in the cool, moist soil deep in marmot or suslik burrows.[26] When a marmot dies in its burrow, they say, it decays and the germs remain, protected from heat and drying, until another marmot comes along and makes its home in the old burrow. If so, plague may not be actively circulating, but it may still be lying in wait. There is plenty of evidence that plague germs can survive in soil under cool, moist conditions. Irkutsk microbiologist Sergei Balakhonov and a team of researchers excavated burrows in Siberia and isolated the plague germ both from sleeping rodents and from the soil itself.[27]

Early Soviet plague control experts did their best to wipe out plague in certain areas by targeting marmots. They poisoned multitudes of the animals with chloropicrin, a poisonous liquid. But each time, after some years, marmots—and plague—always came back. Domaradskij finally argued that the policy was unjustifiably destructive to nature—and that it did not work. So the baiting of marmot burrows ceased.

How plague circulates in nature remains one of the great enigmas of this disease. One plague researcher put it this way: "The ecology of wild rodent plague is most complex; to unravel it is like following the different voices in a Bach fugue, except that in plague the basic design is unknown."[28]

Rats are the principal reservoirs for what is known as domestic plague. It used to be thought that major human epidemics were almost always rat-borne. If we are willing to abandon, at least temporarily, the idea that rats are the necessary and inevitable reservoir for all major human epidemics, we see that human plagues caused by contact with different hosts do not necessarily behave in the same way once the disease enters, from these different portals, the human species.

* * *

Prairie dogs can be endearing: the prairie dog expert John Hoogland found his life's work after driving by a prairie dog colony in Fort Collins, Colorado. Hoogland, an evolutionary biologist, was supposed to study ground squirrels, but he found them singularly charmless. After six weeks trying to observe social behavior among Wyoming ground squirrels, he threw his hands up: "How do you study costs and benefits of group behavior in animals who don't really live in groups?" he asked his wife in disgust.

On April 15, 1974, Hoogland, looking for prairie dog colonies, happened to be driving through the foothills outside Fort Collins. "I drove up over this little rise and saw my first black-tailed prairie dogs. I put on the brakes and I was absolutely mesmerized. I couldn't take my eyes off them. They were fighting, chasing, grooming, sniffing each other, bouncing up and down, giving their territorial calls. I said aloud to myself, I could study these things for the rest of my life."[29]

Hoogland and his team have been watching the prairie dogs for thirty years. Many Westerners don't like the prairie dogs—they encroach on valuable land, they eat crops, and they can carry plague. Massive attempts to exterminate them—including introducing plague-stricken animals into healthy colonies—have gradually given way in many areas to relocation—a process by which prairie dogs are sucked out of their burrows by a device like a giant vacuum cleaner and dropped in another colony better located for human purposes. Hoogland doesn't like this solution particularly—he points out that relocation disrupts the prairie dog colony's social structure completely, and that alien prairie dogs, not genetically related to those in the new colonies, are not going to be able to make a go of it. Also, relocation can introduce plague to a healthy colony. But it's better, from Hoogland's perspective, than slaughtering a whole colony outright.

Hoogland loves the prairie dogs, and he wants to save them from plague, which is a greater threat to their survival than all the murderous depradations of ranchers and developers. He's discovered that infusing Pyraperm, an insectide, into prairie dog burrows stops a plague epidemic in its tracks. This suggests that, despite the "kissing" that prairie dogs do, plague isn't spread pneumonically among them. Also, Hoogland and his associates have never seen the telltale signs of pneumonic plague—bloody froth on the muzzle of a dead prairie dog. The course of the infection is very rapid, Hoogland says. One day a rodent is healthy, darting about, sociable. The next, it seems lethargic, sluggish—and by one or two in the afternoon, the dog is dead.

Plague burns like wildfire through the prairie dog colonies. In a matter of days the entire colony can simply vanish—the dancing and kissing and barking stop abruptly, and the hundreds of earthen mounds that mark burrow entrances are deserted. In the past, some experts thought that pneumonic plague had to be responsible, because the spread was so rapid and the dogs died so fast. But Hoogland's Pyraperm experiment suggests that fleas are at work, moving from dog to dog, spreading the infection with explosive rapidity. If pneumonic transmission were responsible for prairie dog plague, insecticide wouldn't stop the outbreak.

The CDC's Ken Gage has now begun to suspect that prairie dog plague doesn't work like rat plague, that transmission among prairie dogs occurs through mechanical means—contaminated mouth parts—rather than through blockage. "It just moves too fast" for blockage to occur, says Gage.[30]

Two of Hoogland's assistants, his daughter Margaret and a young woman named Jane Jackson, worked with plague-infected prairie dogs in a laboratory one summer while Hoogland himself worked to abort a plague epidemic in one of his study colonies. They were bitten many times by prairie dog fleas, yet neither contracted plague until Jane cut herself while skinning a prairie dog. After five days in the hospital and massive doses of antibiotics, she recovered—but it is interesting that neither Jane Jackson nor Margaret Hoogland got sick from flea bites. Whether this tells us something about the fleas or about the plague strain itself, we cannot yet say.

Prairie dogs are exceptionally sensitive to plague: in less than a century, the prairie dog population in the West has dropped by 98 percent, in large part because of the disease. Plague was apparently introduced into the United States by ships from China carrying infected rats and men.[31] The first anyone knew that plague had come to America was when the body of a dead Chinese sailor was discovered in San Francisco in 1900. From those ships, plague spread into the West, finding new reservoirs among ground squirrels and prairie dogs.

How plague spread into these reservoirs is not hard to discover. As Dr. Walter M. Dickie, secretary of the California State Board of Health, puts it in his 1926 report, "Investigations have repeatedly demonstrated that ground squirrels and rats occupy the same burrows, that rat fleas infest ground squirrels and that ground squirrels infest rats, and that fleas of either species do not object to human society when their host has died."[32] Dickie felt that the contagion could go either way. There is a curious twist to this story: In turn-of-the-century San Francisco, ground squirrels appear to have been something of a delicacy. Done up and presented to the connoisseur as "frog legs," they represented something of a bargain for the chef, as you could get four "frog legs" from each squirrel instead of only two. Tons of squirrels were shot or trapped and shipped into San Francisco and Oakland markets—until in 1908 it was discovered that these "frogs" carried plague.[33]

Both rats and squirrels were the sources, according to Dickie, of numerous outbreaks in California during the years 1900–1925. A large pneumonic outbreak in Los Angeles, which sickened thirty-three and killed thirty-one, was traced to bubonic plague acquired from rats. A strictly enforced quarantine, along with widespread rat trapping, rat proofing, disinfection, medical inspections, and "separation of the rat from his food supply" by sanitary disposal of garbage, among other provisions, seem to have brought this epidemic to a fairly rapid end. This was the greatest, and last, pneumonic outbreak in the United States to date.

Ground squirrels were responsible for one earlier pneumonic outbreak, an August 1919 epidemic in Oakland that took the lives of thirteen people. On August 11 and 13, a man named V. Di Bortoli went squirrel shooting in the foothills of Alameda County. He bagged several squirrels and brought them home to his rooming house in Oakland to cook and eat. On the 15th, he felt ill, and went to his doctor complaining of fever and a pain in his right side. The doctor also found congestion in the right lung, and diagnosed Di Bortoli with influenza. Di Bortoli felt somewhat better, but two days later, a large, tender swelling developed in his neck. The physicians attempted to incise and drain it twice, but nothing was found in the swelling but a bloody fluid. Di Bortoli rapidly grew worse. He was in agonizing pain, pain that even heroin could not relieve. He died the next day.

Five days after Di Bortoli died, another man from the rooming house, a Mr. Toso, came down with pneumonic plague and died three days later. His wife also fell ill, and was diagnosed with influenza pneumonia. She recovered. Within days, the disease had spread to eleven other people, all of whom had been in contact with Di Bortoli, the Tosos, or other victims. All died. But the epidemic was self-limiting, and ended before doctors had even understood that they were dealing with plague.[34] This appears to be the only case on record of pneumonic plague outbreak originating from contact with ground squirrels. This is also the only known chain of pneumonic infection that has involved native American wild rodents—the larger outbreak in Los Angeles was rat-borne.

Prairie dog plague, even when people handle and skin dead dogs, does not seem to ever have "gone pneumonic": cases of secondary pneumonia among people have not spread. When the plague entered the prairie dog populations to the West is not known—but it had to be sometime after it made inroads into the ground squirrel population in California. With so recent an introduction, the prairie dogs have not evolved much resistance, though Hoogland does note that a very few dogs appear to recover from infections. This suggests that plague among prairie dogs does not have to be as virulent as it does among its ancient reservoirs in Central Asia, where the disease has haunted the local marmots and great gerbils for thousands of years. Marmots have evolved resistance over those long years, and the plague germ has evolved greater virulence in marmots to compensate.

Suleimenov and Atshabar say that no plague strains compare in virulence or explosive power to the strains from the reservoirs of Kazakhstan—from marmots and gerbils. "These are the most ancient rodents," says Suleimenov. "They were the initial plague hosts on earth. Research has found out that the plague microbe [originally] circulated in these two rodents."

Suleimenov and Atshabar claim, furthermore, that marmots are the only animals other than humans that suffer from pneumonic plague. "Evidence from the field shows that marmots produce bloody saliva, while other rodents do not."[35] This notion is well supported by an earlier generation of researchers: One, a researcher working in Manchuria, showed in 1912 that marmots could contract pneumonic plague,[36] two others demonstrated in 1917 that marmots could also transmit pneumonic plague directly through coughing to other marmots.[37]

Suleimenov claims that plague germs taken from marmots have a tropism for lung tissue. This means that these plague germs tend to settle in the lungs. Other forms of plague do not seem to show this tropism.[38] "Data from China in 1991 shows that from marmot foci, 45 percent of people who contract plague came down with pneumonic plague," says Suleimenov. "In Vietnam, 98 percent of people with plague develop bubonic plague. Plague in Vietnam is carried by rats!"

The implications of Suleimenov's line of argument are clear, and follow a tradition deeply embedded in Russian thinking about plague based on a long tradition of ecological observation. A Russian scientist named Tumansky argued several decades ago that what separated out plague strains is their adaptation to different rodent hosts. In other words, there are suslik strains, marmot strains, rat strains, gerbil strains—and, though Tumanksy does not apparently mention this, there must have been, during the Black Death, human strains as well. Variations in plague strains result from the pressures of natural selection on the plague germ as it moves into a new host population.

In modern evolutionary terms, each adaptation to a host represents a separate evolutionary struggle between host and pathogen. Plague does not just kill people, it is a great killer of rodents, and these it has slaughtered by the uncounted millions. Sometimes, as with prairie dogs, it seems to be an unequal contest, and it is possible that someday prairie dogs may become extinct. American ground squirrels seem to be doing rather better; they have already proved more resistant to plague.

But, as Suleimenov and Atshabar say, the animals that have probably evolved the greatest resistance are the rodents of Central Asia. In particular, there are four varieties of marmots that play host to the plague germ, and all of them have evolved, over a great span of time, a certain measure of resistance. To overcome that resistance, marmot plague, one way or another, has evolved greater and greater virulence, as Soviet bioweaponeers well understood. They used strains from marmots to make their plague weapon. Major General Nickolai Urakov of Obolensk did not care about plague biology, or about understanding the subtle distinctions among strains. "I only want one strain!" Urakov would bark at his staff. And, for plague, a marmot strain was what he got: a strain that would reliably go pneumonic.

Scientists in the West, in general, have a different perspective than many of their Russian or Kazakh counterparts. In the early 1950s, French researcher R. Devignat, separating out natural plague strains on the basis of arcane biochemical properties,[39] came up with an immensely influential scheme which proposed that all plague strains on earth can be classified into three biological varieties, or biovars. These biovars supposedly represent the three great human pandemics: the Justinian Plague supposedly involved biovar *Antiqua;* the Black Death *Medievalis*, the Third Pandemic *Orientalis. Antiqua* strains are found in Central Asia and parts of Africa; *Medievalis* in Central Asia and Kurdistan, and *Orientalis* in much of the rest of the world, including the United States. So firmly are these categories implanted in the minds of plague specialists, or of anyone who reads a great deal about plague, that it is difficult to shake oneself free of them.

But today, at least in the minds of some scientists, the biovar scheme appears to be breaking down. Since Tumansky, Russian researchers tend to classify plague strains not according to these biochemical properties, but as "host strains": they recognize four main strains, marmot plague, gerbil plague, rat plague, and suslik (ground squirrel) plague. There is also the mysterious, ill-understood *Pestoides* strain, found in voles in the Caucasus region in the south of the former Soviet Union, which some scientists believe to be intermediate between *Yersinia pestis* and its probable ancestor *Yersinia pseudotuberculosis*.[40]

In the United States, a young American scientist, Gary Anderson of the University of California at Berkeley, has been doing work on the genetic fingerprinting of different isolates of *Yersinia pestis*. He feels that Devignat's categories are disintegrating in the face of biochemical analysis, and that only the most recent biovar, *Orientalis*, appears to hold up as a distinct classification. Anderson is a bit of a maverick among American scientists, but, as he learned during a visit to the Anti-Plague Institute Mikrob in Saratov, Russia, his work dovetails quite nicely with Russian host strain theories.[41]

Since no one can point to significant differences in lethality or transmissibility among the three biovars, Devignat's system doesn't provide us much help for our purpose of decoding plague's central mystery. The Russian perspective is much more illuminating. Classifying plague strains not in terms of arcane biochemical properties whose function no one really understands, but on the observable behavior of plague germs in different rodent populations, the host strain analysis casts an entirely new light on how plague strains affect human populations. While very ancient plague strains—those from marmots—are thought by many former Soviet experts, including Atshabar and Suleimenov, to be the most virulent for human beings, it isn't the antiquity of the strains alone that makes them dangerous.[42] *Pestoides*, the Caucasian vole strains, are also thought to be among the oldest strains of all, but they show reduced virulence for guinea pigs in the laboratory; furthermore, according to Russian plague researcher I. L. Martinevsky, there has never been a plague outbreak among human beings traceable to this plague focus.[43]

The older strains have battled it out longer with the hosts in their particular reservoir; in the strains from Central Asian marmots this seems to translate into increased virulence for human beings, though exactly why this should be so is still unknown. Perhaps it is because marmot strains tend to head directly for the lungs, both in marmots and in people.[44]

According to this line of argument, it seems that the answer to the mystery of human plague pandemics lies in the host strains—the portals, one can say—through which the plague germ enters humanity. Of the four major host populations, rats and marmots have historically been the principal and most deadly portals for humanity. But the behavior of plague strains from these two portals once they enter the human species has been quite distinct. To explore this difference further, we must travel far from the laboratory, back to the great pandemics of human history.

# THE WINEPRESS OF GOD

**What is the use of this, when lo!<br>The entire world tottered and came to<br>an end and the duration of generations<br>was shortened? And for whom does<br>the writer write?**

JOHN OF EPHESUS

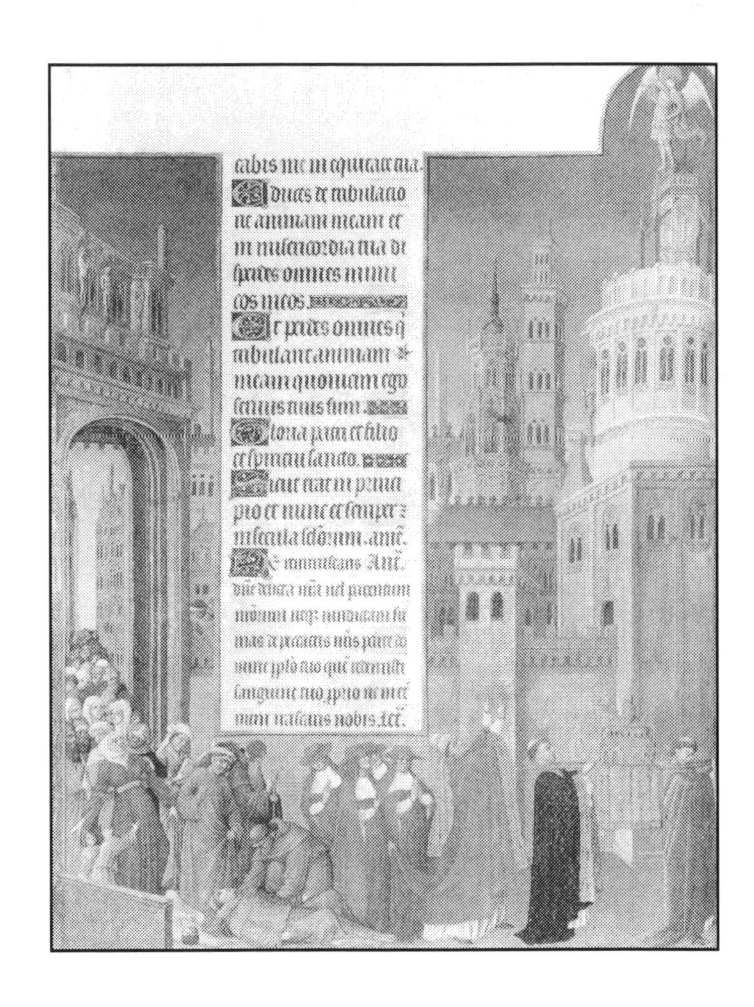

Pope Gregory leading a procession to end the plague, from *Les très Riches Heures.* MUSÉE CONDÉ, CHANTILLY

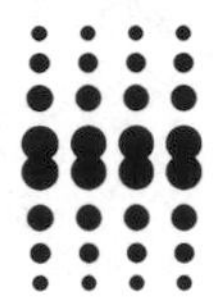

In Samuel V of the Hebrew Bible, the Philistines of Ashdod, who have stolen the Ark of God from the Children of Israel, are struck with an unlikely punishment:

> The hand of the Lord lay heavy upon the Ashdodites, and He wrought havoc among them: He struck Ashdod and its territory with hemorrhoids. When the men of Ashdod saw how matters stood, they said, "The Ark of the God of Israel must not remain with us, for His hand has dealt harshly with us and with our god Dagon."

The Ashdodites consulted with their fellow Philistines, who decided to move the Ark to Gath, another Philistine city. The outcome was not good:

> He struck the people of the city, young and old, so that hemorrhoids broke out among them.

The Philistines tried to move the Ark once again, this time to the city of Ekron. But the Ekronites were having none of it:

> For the panic of death pervaded the whole city, so heavily had the hand of God fallen there; and the men who did not die were stricken with hemorrhoids. The outcry of the city went up to heaven.

Chastened, the Philistines resolved to send the Ark back to the Israelites at once, but their priests and diviners told them that they must pay an indemnity to God as well:

> Five golden hemorrhoids and five golden mice. . . . You shall make figures of your hemorrhoids and of the mice that are ravaging your land; thus you shall honor the God of Israel and perhaps He will lighten the weight of His hand upon you and your gods and your land.[1]

*Hemorrhoids?* Earlier versions of the English-language Bible use the more obscure, and therefore somehow less risible, word *emerods* instead, but the meaning is the same. Smiting people with hemorrhoids is a rather eccentric punishment, especially when you compare it to the ten plagues—the blood, fire, frogs, lice, boils, cattle disease, smiting of the first-born, and so on—inflicted on the Egyptians when the Children of Israel fled Egypt.

Quarts of ink have been spilled by scholars trying to make sense out of these lines. Many have argued that the Hebrew word in question, *ofalim*, doesn't really mean hemorrhoids.[2] It actually means tumors, or swellings. Ancient Hebrew scholars were apparently uneasy with this word, particularly in relation to "secret parts"—it seems to have had some vaguely sexual connotations. To avoid those connotations, later scribes replaced *ofalim* with *t'chorim*, which means hemorrhoids. But "swellings in the secret parts" sounds much less sexual than bubonic, and the association with mice is even more suggestive of plague. The word *akhbar* can mean either mouse or rat, and archaeological evidence clearly indicates that *Rattus rattus*, the black rat that is the principal carrier of plague, could be found in the Levant in that era.[3]

This vague and ominous biblical story is, quite probably, our first literary account of an outbreak of bubonic plague.[4]

The word plague doesn't mean very much when it appears in the ancient texts; then, as now, the word plague—or *plaga*, meaning strike or blow—was used to mean a severe epidemic infection. The plague of Thucydides, described in *The Peloponnesian Wars*, is perhaps the most striking example of a plague that wasn't caused by *Yersinia pestis.* The disease began in Ethiopia above Egypt, passed through Egypt and Libya, and then suddenly fell upon Athens:[5]

> If any man were sick before, his disease turned to this; if not, yet suddenly, without any apparent cause preceding and being in perfect health, they were taken first with an extreme ache in their heads, redness and inflammation of the eyes; and then inwardly, their throats and tongues grew presently bloody and their breath noisome and unsavoury. Upon this followed a sneezing and hoarseness, and not long after the pain, together with a mighty cough, came down into the breast. And when once it was settled in the stomach, it caused vomit; and with great torment came up all manner of bilious purgation that physicians ever named. Most of them had the hickyexe [dry heaves] which brought with it a strong convulsion, and in some ceased quickly but in others was long before it gave over. Their bodies outwardly to the touch were neither very hot nor pale but reddish, livid, and beflowered with little pimples and whelks, but so burned inwardly as not to endure any [but] the lightest clothes or linen garment to be upon them nor anything but mere nakedness, but rather most willingly to have cast themselves into the cold water. And many of them that were not looked to, possessed with insatiate thirst, ran unto the wells, and to drink much or little was indifferent, still being from ease and power to sleep as far as ever. As long as the disease was at its height, their bodies wasted not but resisted the torment beyond all expectation; insomuch as the most of them either died of their inward burning in nine or seven days whilst they had yet strength, or, if they escaped that, then the disease falling down into their bellies and causing there great exulerations and immoderate looseness, they died many of them afterwards through weakness.[6]

No one has determined what this disease outbreak was, though smallpox, measles, and even (improbably) Ebola have been proposed as candidates.

An ancient attack of what appears to be plague was recorded by a physician known as Rufus of Ephesus, who lived during the reign of the Emperor Trajan in the second century B.C.E. Only a fragment of Rufus's work survives, in the compilation of the fourth century C.E. writer Oribasius.

> The buboes that one calls pestilential are very acute and often cause death. It is especially in Lybia, Egypt, and Syria that they are seen to occur. Dionysius Curtius the Humpback has referred to these buboes. Dioscorides and Posidonius have referred to them at length in their treatise on the plague which in their time raged in Lybia, and they have said that it was accompanied by an acute fever, intense fever, intense pain, perturbation of the whole body, delirium, eruption of large buboes hard and without suppuration, developing not only in the usual places but also in the popliteal space and elbow, although in general such inflammations do not form in those places.[7]

Rufus also says that "one can foresee a plague which approaches by paying attention to the bad condition which the seasons present; to the manner of living less profitable for health, and to the death of animals which precedes its invasion."[8]

Plague seems to have broken out again and again in the ancient world, always—so far as we can tell—centered around the Levant and North Africa. These accounts are remarkably consistent with what we know of rat-borne plague. But these outbreaks, lethal and devastating though they must have been, did not spread across the known world—they never became pandemics. We don't know all the factors necessary to turn sporadic outbreaks into epidemics like the one that "raged in Lybia" or that sent a cry up to heaven from

Ashdod, though—odd as it is—Rufus's three criteria sum up matters quite well.

But to become pandemic, plague requires more than rats, fleas, or other means of contagion. War and commerce are the handmaids of plague contagion. Without the large-scale movement of goods and of people, the massive explosions of the three historical pandemics—the Justinian Plague, the Black Death, the Third Pandemic—could never have taken place.

It was the world of the Emperor Justinian that set this ancient apocalypse into motion.

* * *

> Caesar I was, Justinian I am.
> By the will of the First Love, which I now feel,
> I pruned the law of waste, excess and sham. . . .
>
> As soon as I came to walk in the True Faith's way,
> God's grace moved all my heart to my great work;
> And to it I gave myself without delay.[9]

In 542, the year the plague came to Constantinople, that city had long been the seat of the Roman Empire. Rome, and Italy, had fallen, first to Alaric the Visigoth in 410, then to Odoacer the Scirian; by 542, Byzantines and Ostrogoths were still contending over it. North Africa had been ruled, until recently, by the Vandals, and periodic incursions by Berber tribesmen had devastated that rich countryside. In the east, along the Persian frontiers, raids and counterraids were constant, and a major war depleted much of the empire's resources. And beyond all that, barbarian hordes continued to press upon the empire, wave upon wave: Franks, Gepids, Langobards, Slavs, Utrigurs, Kotrigurs, Avars. All these endless battles and wars cost the empire men and money. Religious struggles, too, rippled across the empire from east to west: struggles for supremacy among the various Christian factions, attempts to root out heresy by execution or by decree; skirmishes with restless minorities like the Samaritans and Jews. There were, also, the perturbations caused by wandering holy men, or of filthy hermits living in caves, in monasteries, or atop the sixty-foot pillars where they spent their lives, intervening between God and the multitudes who came to them, begging for their wisdom and their assistance.

The emperor of all this unrest was Justinian, a Thracian peasant's son who had been chosen by the Emperor Justin, his uncle, to succeed him. Justin himself, once a mere count of the Excubitors in the Imperial Bodyguard, was sixty-eight when he ascended the throne. His predecessor, Anastasius, had died without an heir, but the illiterate and uncouth Justin had the support of his soldiers, who pushed aside Justin's competitors. His nephew Peter, also a guardsman and a senator, became his heir apparent; Peter renamed himself Justinian in his uncle's honor, and soon, through his intelligence, energy, and political astuteness, became indispensable—first the de facto ruler of the empire, and, after Justin's demise in 528, emperor himself.

But Justinian did not rule alone. Nearly as powerful as the emperor, and far more to be feared, was his wife, Theodora. This remarkable woman, the Eva Perón of her time, had a delicate beauty, the courage of a soldier, and an adamantine will. She came from the lowest levels of Constantinople society: her father was a bear-keeper in the Hippodrome, the giant arena where sports and political conflicts were often played out, where the twin factions of the Blues and the Greens, rival gangs that overran Constantinople at the time, had their seat. Theodora's father belonged to the Greens; after he died, her mother found herself facing starvation. She married again quickly, but her new husband had no position. She sent her three daughters—Theodora, the middle child, was not more than five years old—into the Hippodrome to appeal to the factions to give her new husband the vacated bear-keeper's post. The Greens jeered at the pleading children, but the Blues showed them compassion. Theodora never forgot this compassion: in the future, Theodora and Justinian would support the Blues against the Greens, adding to the turmoil of the city.

A few years later, Theodora followed her sister Comito onto the stage at the Hippodrome. She became a comedienne, but of a rather peculiar type: her salacious performances, and her exploits as a prostitute, became notorious throughout the city. Among other acts, she would lie down naked on stage, have servants adorn her private parts with grains of corn, and set a goose to pluck them off, to the evident amusement of the crowd. Edward Gibbon reports that "her charity was *universal*" and that that her "murmurs, her pleasures, and her arts must be veiled in the obscurity of a learned language."[10] He proceeds to specify them in Greek and Latin.

Eventually she took up with a wealthy merchant from North Africa and left Constantinople with him. The liaison ended; adrift, she wandered from city to city, probably supporting herself through prostitution, until she ended up in Egypt. There she fell under the influence of Severinus, a clergyman, and cast off her way of life. She returned to Constantinople and supported herself quietly as a seamstress until—probably through the Blue faction—she managed to meet and enthrall Justinian. The abstemious thirty-six-year-old senator was overcome by her beauty and experience; he persuaded his uncle Justin to change a long-standing law forbidding senators from marrying actresses. Not long after their marriage, the Thracian peasant and the actress-prostitute donned the royal purple of the empire. They remained devoted to each other; Justinian learned to rely on the courage and formidable intelligence of his consort, who became his chief confidante and advisor. Together, they became the greatest rulers of the late Roman Empire, which was much diminished from its earlier glories, and which they tried hard to restore. Much of their work would be undone when the plague came to Byzantium.

Justinian began his reign by commissioning a group of jurists to codify a body of laws for his empire: this massive work was known as the Justinian Code, which, in the words of one scholar, "aimed to include any valid law of any date [in the history of the empire] and to eliminate anything that had been superseded."[11] The code was a remarkable success, and added significantly to his prestige. The emperor also intervened in religious matters, adjudicating freely among the many warring Christian factions of the time. Intolerant of pagans and heretics, he closed down the once glorious pagan Academy in Athens, fought the Samaritans, taxed the Jews, and executed the pacifist but obdurate Manichaeans wherever he could find them.

But Justinian's lifelong, overriding passion was the resurrection of the Roman Empire. To do so he needed to wrest Italy from the Ostrogoths, and North Africa from the Vandals, though he had little hope of Gaul or Britain; he also needed to secure the eastern frontier, the empire's weakest point. He wanted to restore the empire's suzerainty over these lost provinces—and he wanted their tax revenues as well. Justinian and Theodora, arrivistes that they were, had a hunger for luxury, for a splendid imperial court, for the greatest buildings, churches, and public works they could erect. That took gold, and taxes, and trade, more than his shrunken dominions could provide. Justinian lost no time in setting about to regain those portions of the old Roman Empire now in barbarian hands.

Shortly after his accession, Justinian sent an army to overthrow the Kingdom of the Bosphorus (now Crimea), which he annexed. His generals also defeated the Persians in several battles; Justinian strengthened his eastern borders with a series of fortresses, and eventually signed a peace treaty with the Persian king.

The eastern border quiet for a time, Justinian turned his attention to North Africa and the Kingdom of the Vandals. The Vandal king Gelimer had deposed and imprisoned his predecessor, Hilderic, who had been friendly toward Constantinople; Justinian used

Gelimer's act as an excuse to launch a massive attack. North Africa was particularly important because it was the grain basket of the empire: wheat from North Africa was shipped to Byzantium and stored in a massive warehouse Justinian had built for the purpose—a fact not without significance in the history of the plague.

Gelimer, distracted by a revolt in the province of Sardinia, was taken unawares by the empire's forces, which sailed in, landed, and quickly overran the country. The citizens of Carthage, Roman citizens of old who hated their Vandal overlords, opened the gates of the city to them. After a series of decisive victories, the Vandals were routed, and sovereignty over North Africa was restored. But the Berbers of North Africa were not so easily pacified as the Vandals; they continued their raids, and the Byzantine army had to remain in place in Africa, restive, underpaid, and a drain on Justinian's coffers. Nevertheless, prosperity returned to North Africa, which became again one of the richest provinces of the empire.

Justinian next turned to Italy, which until 526 had been under the relatively tolerant and enlightened rule of the Ostrogoth king Theodoric.[12] His heir, Eutharic, however, died before he did, and Theodoric's death in 526 left his eight-year-old grandson, Athalaric, on the throne, with his mother, Theodoric's daughter, the beautiful, strong-minded Amalasuntha, as regent. The Goths hated Amalasuntha for her strength, her attempts to train her son in literature and philosophy, and her leanings toward Justinian and Constantinople. She imagined that she could unite the Goths and the Roman Empire into a single civilized entity, but the Goths would have none of it. Despite her influence and her attempts to have him schooled, her son was a weak-minded boy who ran crying to the Goths at court after his mother once struck him. They took over his education, mostly in fighting, and introduced him to drink and women. By sixteen he was dead of dissipation. Amalasuntha tried to take power into her own hands, but the Goths did not want a female ruler, and she knew her life was in danger. She appealed to

Justinian for help, but she was imprisoned and then killed by the order of her cousin Theodahad, a man she had invited to serve as co-ruler with her. Theodahad had hated her for years, ever since she forced him to return estates in Tuscany he had extorted from local landowners. He imprisoned Amalasuntha on a remote island, and had her strangled in her bath. This ignominious murder of a king's daughter—in which Theodora, jealous of Amalasuntha's beauty, learning, and wit, apparently connived[13]—gave Justinian the pretext he needed to invade Italy and take it back under imperial control.

The Gothic War raged for years, under one Gothic leader after another. Justinian's great general Belisarius took Rome, but he could not hold it. Still, by 541, after a long struggle, most of Italy had been pacified.

But the wars in the west left the eastern border with Persia vulnerable. In 532, Justinian had signed a "Perpetual Peace" agreement with Persia, but the manpower drain for Justinian's western wars had given the Persian king Khusrau an opening to renew plundering expeditions against the empire. Khusrau himself led a large host into Syria, then under imperial control, to sack several large towns, including the great city of Antioch. They burned most of Antioch, and held thousands of the survivors for ransom, a ransom Justinian could not or would not pay. There were other Persian provocations—in 541, Khusrau moved against Lazica, a Byzantine protectorate on the Black Sea, and sent his allies the Huns to strike Armenia.

Not long after, General Belisarius, removed by Justinian from Italy, marched on Persia itself. This was the beginning of the long, debilitating Persian War: Justinian now had to rebuild Antioch and other sacked cities and rebuild many destroyed fortifications before he could take on the Persians in earnest.

Justinian had no intention of conquering Persia and adding it to his dominion: he wanted to restore and protect the empire—and to build up his treasury. His goals in the struggle with Persia were essentially commercial. The war with Persia was a war over silk; the Persians held a monopoly on the overland silk trade from China, and bought up all the silk in the Indian markets.[14]

Still, by 541, commerce in the Byzantine Empire was now flourishing. Constantinople, with its enviable maritime position and splendid ports, was ideally located to be the commercial hub of the empire.[15] As one scholar puts it, "Never was a capital situated more advantageously than Constantinople to become the foremost commercial city of the world. Constantinople was not merely the market of the Black Sea and the Aegean archipelago; the Levant was her dominion; Syria and Egypt paid tribute; her commercial interests radiated as far as China and India."[16]

The tribute that Egypt paid was tribute in grain—about 240 metric tons per year.[17] This vast amount of wheat was brought in to Constantinople for the bread dole—central to the extensive social welfare system in the city. Justinian taxed his subjects mercilessly for his great building projects, the imperial magnificence of his palace, and his endless wars—but he also fed them, though their diet was minimal: bread, vegetables, a little wine.[18] Like the Romans before them, the populace of Constantinople lived on bread and circuses—the bread of Egypt, the circus of the Hippodrome. To satisfy the insatiable need of his city for grain—over half a million lived in Constantinople—Justinian built an enormous storehouse at the mouth of the Hellespont, 90 feet by 280 feet; and "ineffably" high.[19] How much grain—and how many rats—such a structure could contain, it is difficult to imagine. There were also at least four other granaries dating from earlier times, and other cities such as Alexandria had their massive storehouses as well. These massive structures became breeding grounds for plague. A single breeding pair of stowaway rats among the grain could produce one thousand offspring *per year*. Granaries form excellent flea nurseries—the Oriental rat flea, *Xenopsylla cheopis*, does not feed on blood in its larval form, but rather on the debris from grain or rice.[20] Justinian's granaries must have supported vast and ever-growing populations of rats and fleas.

Egyptian merchants traded in other goods as well, including woven silk and fine linen, also havens for infected fleas.[21]

The harbors of the empire must have been crowded, noisy places: the knocking of hull against wharf as the ships were brought in, the shouts of men, the unloading of bale after bale of grain, of silk, of linen; the rattle of carts carrying the goods away, the scurrying of rats off the ships into the port.

By 542, Justinian had resurrected most of his empire. Its commercial networks spanned the known world, and its armies spanned Europe and the Mediterranean. It was not so much the movements of men that carried the plague, for this plague was not a human disease. It was not war, but peace that brought the plague to Constantinople: peace, and the prosperity that comes with reinvigorated commerce. The ships, the grain, the linen and embroidered silk brought from Egypt to cities throughout the empire: these were tinder waiting for the spark of pestilence.

* * *

The catastrophe that fell upon Justinian's reborn Roman Empire was unprecedented in human history. Local plagues were devastating enough, killing tens of thousands. Still, they burned out and disappeared. We do not know whether the plague of Athens was a visitation of measles or smallpox or some other disease unknown today, which raged briefly and then vanished off the earth. The plague outbreaks mentioned by Rufus of Ephesus were almost certainly true plague, but restricted to relatively small areas—Egypt, Syria, Libya.

Justinian's plague was the first true plague pandemic. Procopius, who hated Justinian, blamed the emperor directly for the plague.

He suggested in his malicious *Secret History*, unpublished in his lifetime, that the emperor was a demon and either created the plague himself or was punished by the deity for his wickedness. Procopius was more right than he knew. Justinian had not created the disease, but he had created the pandemic, which followed the movements of men and goods in Justinian's resurrected empire. Without that empire, the bread dole, the huge shipments of grain and cloth from Africa, it is difficult to imagine how the First Pandemic could ever have erupted.

"During these times there was a pestilence, by which the whole human race came near to being annihilated," says Procopius.[22] How many people died it is impossible to know; Gibbon suggested, based on some calculations from Procopius, that a figure of 100 million dead was "not unreasonable": from this shallow basis the number 100 million has crept into the literature and can be found in many current scientific papers.[23] But there is actually no direct, unassailable physical evidence of the plague's footprints across the known world, nor of the actual damage that it did. For the Second Pandemic, the Black Death, there are plague pits and cemeteries, deserted villages, and hundreds of sources from Europe, the Middle East, and Asia. An analysis of these sources led the nineteenth-century scholar J. F. B. Hecker to propose a figure of 25 million dead from the Black Death in Europe alone. Given that the Justinian Plague had far less of a reach than the Black Death, the death toll must have been considerably less than 25 million.

For the Justinian Plague, we have only a half-dozen primary sources of varying clarity and usefulness, and bare mentions in a handful of other texts. These accounts stretch across the length of the empire and over a span of 280 years. Of these, Procopius's account in his scrupulous *History of the Wars* is much the clearest and best: the ur-text, as it were, on the First Pandemic, which, as he said, "embraced the entire world and blighted the lives of all men."

The plague, Procopius tells us, began in Pelusium, a Mediterranean port on the eastern part of the Nile Delta. (Another source, the ecclesiastical historian Evagrius, who himself contracted the plague in 542 when he was six years old, claims that the disease actually first broke out in Ethiopia, or Axum, and from there spread upward to Egypt, a scenario that is more likely.) Procopius continues:

> Then it divided and moved in one direction towards Alexandria and the rest of Aegypt, and in the other direction it came to Palestine on the borders of Aegypt; and from there it spread over the whole world, always moving forward and travelling at times favourable to it.
>
> For it seemed to move by fixed arrangement, and to tarry for a specified time in each country, casting its blight slightingly upon none, but spreading in either direction right out to the ends of the world, as if fearing lest some corner of the earth might escape it. . . . And this disease always took its start from the coast, and from there went up to the interior.

That the disease "took its start from the coast" suggests that plague followed the trade routes. We know from archaeological research that the presence of black rats in the Mediterranean littoral dates at least from the neolithic; we also know that black rats are not by nature nomadic. They either stay within a few hundred yards of where they were born, or they may inhabit ships, debarking at each port; they can use any conveyance, including saddlebags on camels, or a wheeled cart filled with grain, but they don't, by themselves, journey cross-country.[24] The black rat "has never been found further than two hundred meters away from a building and lives mostly in granaries and on ships, where it is almost invariably found. It never moves from one village to another or from one port to another, except when passively transported."[25]

Procopius describes the disease in exacting detail, as Thucydides, his model, had done for the Athenian plague: "I shall proceed to tell where this disease originated and the manner in which it destroyed men." Thucydides he wasn't, though; he exhibits a modest credulity more typical of his religiously excitable era than the dry rationality of his Athenian forebear:

> Apparitions of supernatural beings in human guise of every description were seen by many persons, and those who encountered them thought that they were struck by the man they had met in this or that part of the body, as it happened, and immediately upon seeing this apparition they were seized also by the disease.

Others died after seeing visions in dreams of similar apparitions, or heard voices foretelling that they were written down in the number of the doomed. But the majority, Procopius concedes, were struck by the plague without any advance warning beyond a slight fever.

> And the body shewed no change from its previous colour, nor was it hot as might be expected when attacked by a fever, nor indeed did any inflammation set in, but the fever was of such a languid sort from its commencement and up till evening that neither to the sick themselves nor to a physician who touched them would it afford any suspicion of danger.

This is the way that the plague germ kills: silently, subtly, it steals through the host, preventing inflammation, which would alert the immune system that the body is under attack.

> But on the same day in some cases, in others on the following day . . . a bubonic swelling developed; and this took place not only in the particular part of the body which is called "boubon" [the groin], that is, below the abdomen, but also inside the armpit, and

> in some cases also beside the ears, and at different points on the thighs.

These, of course, are the familiar places where buboes form: the femoral (on the thigh), inguinal (the groin), axillary (the armpit), and cervical (neck) buboes familiar to plague clinicians today.

Everyone who fell ill, according to Procopius, came down with these buboes—but from there on the disease followed a wildly variable course:

> For there ensued with some a deep coma, with others a violent delirium, and in either case they suffered the characteristic symptoms of the disease. For those who were under the spell of the coma forgot all those who were familiar to them and seemed to be sleeping constantly. . . . But those who were seized with delirium suffered from insomnia and were victims of a distorted imagination; for they suspected that men were coming upon them to destroy them, and they would become excited and rush off in flight, crying out at the top of their voices.

This is remarkably similar to a twentieth-century account from Manchuria.

> The patient becomes weaker. His mind wanders. He is apathetic. Cerebration is impaired. He cannot control his speech. Instead of apathy, there may be wild delirium . . . some patients show a marked *Wandertrieb*, attempting to run away from the hospital. Symptoms of air hunger are frequently displayed, the sufferers wanting to leave the room in order to sit or lie in the courtyard. . . . In a minority there may be marked agony accompanying oedema of the lungs, or low muttering delirium developing into coma may close the scene.[26]

The picture in Procopius's account is of classic bubonic plague. Furthermore, Procopius insists that the disease was not contagious:

> Neither physicians nor other persons were found to contract this malady through contact with the sick or with the dead, for many who were constantly engaged either in burying or in attending those in no way connected with them held out in the performance of this service beyond all expectations, while with many others the disease came on without warning and they died straightaway.

Procopius also tells how physicians, at a loss to explain or understand the disease, opened the buboes after death, and found that "a strange sort of carbuncle . . . had grown inside them." Some people broke out in black pustules "about as large as a lentil"; these people "did not survive even one day, but all succumbed immediately." This rapid death, along with the black pustules, suggests septicemic plague. Procopius also describes sudden vomiting of blood that "straightway brought death."

Finally, Procopius points out that in those patients where the buboes swelled to an unusual size and burst, the patient usually survived, "for clearly the acute condition of the carbuncle had found relief in this direction."

The disease raged in Constantinople for four months. At first the death rate was just a little higher than normal, but it rose gradually, and eventually reached *five thousand a day.* At first, people attended diligently to their own dead, but as the death toll mounted, "confusion and disorder everywhere became complete." Bodies lay about the streets, and in the houses, unburied—there was no one to bury them. Slaves lost their masters, and masters their slaves; many houses stood empty. Justinian took charge, paying for the disposal of bodies from his treasury, and assigning soldiers from the palace to help bury the dead. The Blues and Greens, too, their enmities forgotten, marched together to dispose of the bodies: these young toughs "learned respectability for a season by sheer necessity."

Eventually all the tombs already built were filled, so the soldiers dug trenches everywhere possible throughout the city, and tossed the bodies in. But that expedient, too, failed, as there was nowhere else to dig. Next the soldiers tried to dispose of the corpses in the towers of fortifications in one particular quarter: they tore off the roofs and threw in the dead without care, without order, without funeral rites.

Disposing of the dead aboveground had its disadvantages: a foul stench blew across the city, to the anguish of its remaining inhabitants. Eventually the towers were exhausted, and still more corpses lay about in the streets. Finally, people simply carried the bodies down to the sea and tossed them in, or onto skiffs, in heaps "to be conveyed where it might chance."

Justinian himself fell ill, and nearly died, sending the palace into turmoil: after a long convalescence, however, he recovered.

The death rate remained high for three months. We cannot know how many died, but it must have been tens of thousands out of a city of some 300,000 people. How Procopius arrived at his figure of five thousand dead a day at the plague's height we cannot know. But it is not unreasonable to assume that 20 percent of the city died.[27]

Another account comes from the ecclesiastical historian Evagrius Scholasticus, who was six when the plague came to Antioch in 542, and who nearly died of it. The plague was to shadow all Evagrius's life: when he wrote his history, he had already lived through four epidemic waves, in which he lost much of his family, including his wife and many children. Evagrius was a man of some distinction, a lawyer and an author, awarded titles by the Emperors Maurice and Tiberius for his literary work.[28] Unlike Procopius,

Evagrius was a man of Orthodox faith. Only the plague caused him to waver in that faith, and then only briefly.

He offers a rather vague account of the spread of the disease:

> And the ways in which it was passed on were various and unaccountable. For some were destroyed merely by being and living together, others too merely by touching, others again when inside their bedchamber, and others in the public square. And some who have fled from diseased cities have remained unaffected, while passing on the disease to those who were not sick. Others have not caught it at all, even though they associated with many who were sick, and touched many not only who were sick, but even after their death. Others who were indeed eager to perish because of the utter destruction of their children or household, and for this reason made a point of keeping company with the sick, nevertheless were not afflicted, as if the disease was contending against their wish.[29]

Is Evagrius, in the first several lines, describing contagion? It is difficult to say; pneumonic plague doesn't appear to be part of the picture, but the account of people catching the disease within a sick person's bedchamber could suggest transmission by human fleas.

A devout Christian who believed that the disease was the sign of God's wrath at sinners, Evagrius could not understand why he had lost so many family members, while "this never happened to pagans with many children." He went to see St. Symeon the Younger, a famous holy man, who had been living on a pillar since childhood (he got his second set of teeth on the column, Evagrius tells us). At that time, St. Symeon lived on top of the Miraculous Mountain in Syria: a whole monastic colony and a great church were eventually built around his pillar, and the ruins can be seen to this day.[30] The holy man, filthy, bearded, and wild-eyed, stared down from his sixty-eight-foot pillar at Evagrius and divined his thoughts: his grief at the loss of his children, his anger at God for sparing the pagans. St. Symeon told Evagrius that such thoughts were displeasing to God, and that he must put them away. Evagrius rushed up the holy mountain to beg the saint for forgiveness.[31]

* * *

John of Ephesus, whose writings survive only in fragments, never doubted that the plague was God's work:

> The blessed prophet Jeremiah . . . would cry and lament not over the destruction of one single city . . . but over many cities which Wrath had, as it were turned into wine presses—[it treaded][32] and squeezed inside them all the inhabitants without mercy, as if they were ripe grapes; over the whole earth, because the (divine) decree was issued . . . it cut down and threw people of various statures, distinctions, and ranks together and [without exception]; over corpses that burst open and stank in the streets with no one to bury them; over large and [small] houses, pleasant and attractive, that suddenly became graves for their inhabitants, in which both servants and masters fell suddenly together. . . .
>
> Now that I, the miserable one, wanted to include these events in a record of history, my thoughts were blocked by many fears, . . . Mouths and tongues are all deficient in describing it. Besides, even if they were able to describe one portion of it, what is the use of this, when lo! the entire world tottered and came to an end and the duration of generations was shortened? And for whom does the writer write?

John says of future generations, "Will they learn through the punishment of us, the miserable ones, and will they be saved from the Wrath of the present and from the punishment of the future?"

The dreams and waking visions of Procopius have become, in John's text, nightmare specters, which haunt the seas in boats of copper, pushed by headless people with copper poles. At night the boats burn with a spectral fire. Some of these headless black figures are seen heading toward Gaza and Ashkelon on the coast of Palestine; people looking on them nearly die of fright. In the countries passed by the burning boats, the plague begins.

The houses of the wealthy are left standing empty; those who try to enter and take their riches die at once. In one Palestinian city, men behold angels who turn out to be devils in disguise; they convince the people to worship a bronze idol to avert the plague. An enraged God strikes all of them down for their misplaced sacrifices: *"All of them were destroyed because they did not remember the name of the Lord."*

John was in Palestine when the epidemic of 542 was at its height; at that point he and his companions fled to Mesopotamia, which cannot have been much better. "Day after day we too used to knock at the door of the grave along with everyone else."

John eventually came to Constantinople, but he could not escape the scourge: "Thus when the punishment, in which God's abundant kindness and grace were really visible, reached this city, though it was horrible, powerful and intense, we ought to call it not only a threatening sign and wrath but also a sign of mercy and a call for repentance."

At this point, for the first time in his narrative, lamentation and legend yield to observation, and John of Ephesus gives us a clearer picture of the actual spread of the plague:

> When it invaded a city or a village, it fell furiously and quickly on it as a reaper, on its suburbs. . . . And thus, after becoming firmly rooted, it moved along slowly. This is what happened to this city [Constantinople], after it learned about the progress of the plague through hearsay from everywhere, during a period of one or two years.
>
> . . . it started with vigour first with the masses of poor people

> who were cast away in the streets. Sometimes five, seven, twelve, and up to sixteen thousand persons among them were removed in a day. Since it was still the beginning, men used to stand in harbours, straits and gates, counting their numbers.

John says that the counting stopped after 230,000. After that, he says, the dead were removed without counting. He also mentions that both those who died and who survived infection had been "stricken with a blow"—the swelling of their groins, or, in Syriac, "the disease of eruptions."

Animals died as well: "not only . . . the domestic ones, but also . . . those of the steppe, including even the reptiles of the earth. It was possible to see cats, dogs, other animals, and even mice, whose groins were swollen and were cast away and dead."

There is another, mysterious sign John mentions: three pock-shaped black marks in a person's palm, the inevitable sign of death.

The dead often perished suddenly, almost without warning—a description we see again and again in the accounts of the Black Death: people would die at work, in their daily bath, in the marketplace: buyers and sellers would converse, pass over money and goods, and then, without warning, fall over right there in the street.

John speaks, too, of the death ships mentioned by Procopius: ships filled with corpses that were dumped into the straits, which would then return for more of their frightful cargo. Great mounds of corpses were stacked like wood all along the seashore; from those rotting, bursting corpses putrefaction slid down into the water, a slow stream of corruption.

> Lovable children were mixed together and cast away; those who cast them in ships seized them and tossed them from afar in great terror; handsome and cheerful boys, who turned dark, were thrown, head down, beneath each other, and made into an object of terror; noble and chaste women, dignified and respectful, sat in

> their private rooms, their mouths swollen, exposed and wide open; they were piled up in horrible heaps.

In the plague pits where the corpses were thrown the winepress of God became a winepress of humanity: those who buried them would trample on the corpses to push them further down, into a noisome heap of disintegration, to make room for more to come. The fastidious Procopius averts his eyes from the horror of the plague pits—he never mentions them—but John does not flinch from what was surely one of the ghastliest sights in human history.

* * *

Gregory of Tours, a Gallo-Roman bishop who would later be canonized, also knew plague was the wrath of God for the sins of men. Gregory was born in 539, and he witnessed the second major plague wave, whose ripples spread from the Mediterranean well into what is now France. His record of the plague outbreak in Marseilles is meticulous and restrained:

> At this time [588 C.E.] it was reported that Marseilles was suffering from a severe epidemic of swelling in the groin and that this disease had quickly spread to Saint-Symphorien-d'Ozon, a village near Lyons. . . . A ship from Spain put into port with the usual kind of cargo, unfortunately also bringing with it the source of this infection. Quite a few of the townsfolk purchased objects from the cargo and in less than no time a house in which eight people lived was left completely deserted, all the inhabitants having caught the disease. The infection did not spread through the residential quarter immediately. Some time passed and then, like a cornfield set alight, the entire town was suddenly ablaze with the pestilence.[33]

This is exactly the pattern we would expect from rat- and rat-flea-borne plague. Perhaps the initial infections were caused by infected fleas infesting the cargo—this might account for the first wave of disease. It takes time for the plague to manifest itself widely in the residential quarter. Infected rats, scurrying off the ship into the port, first have to die and spread their fleas to the native rats of the harbor. But then, after an interval, the plague explodes.[34]

Gregory goes on to say that after two months the plague burned out, and many people who had fled Marseilles drifted back home again—only to find that the plague burst out once more. "All who had come back died," he tells us, and, furthermore, that Marseilles suffered several such epidemics. Gregory could not know that this would be true for over a millennium: plague last struck Marseilles in 1720. The suffering of this great port city from plague is no coincidence—it is a natural conduit for plague from the sea into the lands around the Mediterranean.

Gregory is also the source of a curious story about an outbreak in Rome, which became part of the religious mythology sprung up around the plague. In 589, Gregory's deacon Agiulf was in Rome, and witnessed the River Tiber overflowing its banks, so that many ancient churches and the papal granaries—containing some seven thousand bushels of wheat—were destroyed. Then a great school of water snakes swam down the river to the sea; in their midst, Agiulf reported to Gregory, was a vast dragon "as big as a tree trunk." All these creatures were destroyed "by the great salt wave"—and their bodies washed up along the riverbanks, rotting and stinking. Soon afterward, a great plague epidemic, which caused swellings in the groin, broke out, claiming Pope Pelagius as its very first victim. The next pope was another Gregory, sprung of a wealthy senatorial family, who had already established and supplied six monasteries and sold the rest of his goods and lands to give to the poor. He did not want to be pope, and had to be induced by his fellows to accept the papacy. Famed for his learning and his abstemiousness, he was a frail man whose frequent fasts had rendered him so thin that "his weakened stomach could scarce support his frame." But he became the first great pope of the early Middle Ages, and was to be known as Gregory the Great. Immediately after his ascension, he called for masses and penitential processions to avert the plague: he told the people of Rome that "our present trial must pave the way for our conversion" and promised that "When He sees the way we ourselves condemn our own sins, the stern Judge may acquit us of this sentence of damnation which he has proposed for us."

The stern Judge may have been mollified by Pope Gregory's processions: the plague ceased, and the pope went down in Church history for having stopped the plague in Rome.

What connection there may have been between the plague outbreak and the destruction of the wheat in the granaries, with their undoubtedly vast population of drowned or hungry rats, Gregory of Tours does not say.

* * *

Some of the most powerful and dramatic lines written about the Justinian Plague were written by Paul the Deacon, a Lombard in northern Italy, nearly two centuries after it had passed:

> The flocks remained alone in the pastures with no shepherd at hand. You might see villas or fortified places lately filled with crowds of men, and on the next day, all had departed and everything was in utter silence. Sons fled, leaving the corpses of their parents unburied; parents forgetful of their duty abandoned their children in raging fever. If by chance long-standing affection constrained any one to bury his near relative, he himself remained unburied, and while he was performing the funeral rites he perished; while he offered obsequies to the dead, his own corpse remained without obsequies. You might see the world brought back to its ancient silence: no voice in the field; no whistling of shepherds; no ly-

> ing in wait of wild beasts among the cattle; no harm to domestic fowls. The crops, outliving the time of the harvest, awaited the reaper untouched; the vineyard with its fallen leaves and its shining grapes remained undisturbed while winter came on; a trumpet as of warriors resounded through the hours of the night and day; something like the murmur of an army was heard by many; there were no footsteps of passers-by, no murderer was seen, yet the corpses of the dead were more than the eye could discern; pastoral places had been turned into a sepulchre for men, and human habitations had become places of refuge for wild beasts.[35]

Dramatic these lines are, and terrifying, like the laments for Jerusalem in the Bible. They are perhaps a little too much like laments in the Bible—and too much like Thucydides as well. Thucydides makes it clear that the Athenian plague was extremely contagious: "For at first neither were the physicians able to cure it . . . but died fastest themselves, as being the men that most approached the sick." And also: "For if men forebore to visit [the sick] from fear, then they died forlorn . . . if they forebore not, then they died themselves, and principally the honestest men."[36]

In Paul's account we hear an echo of Thucydides—not his style, not his precision, but an echo of the description itself, and of the fear of contagion and abandonment.

Among these narratives of Justinian's Plague, it is only in Paul's history that the fear of contagion surfaces. Despite the horror and the devastation of the Justinian Plague, despite the palpable anguish of Evagrius and John of Ephesus, none of the eyewitnesses ever mention people contracting the disease one from another.[37] Would anyone have trampled on the oozing bodies of the dead if they feared catching the disease from the corpses?

The contagion in Paul's work, apparently, is literary contagion only. Paul wrote from historical memory and literary precedent.

Even the description of a bubo as like a date or a nut could only have been written by someone who had never seen a bubo for himself. Since the plague outbreak he is describing here took place nearly two hundred years earlier, it seems gratuitous to point out that we cannot take this account as much more than poetry or fable. And yet this haunting, poetic passage is quoted even in scientific writing as one of our principal sources for the Justinian Plague.[38]

Paul also repeats Gregory of Tours's tale of the flooding of the Tiber, with its attendant water snakes and dragon, and of the plague epidemic that broke out afterward in Rome, ending only when Pope Gregory called forth his penitential processions. The story made its way into one of the most influential books of the Middle Ages, *The Golden Legend* of Jacobus de Voragine. This medieval Latin collection of saints' lives and legends was compiled in the mid-thirteenth century, and was read throughout Europe—over five hundred manuscript copies of it have been found.[39] Perhaps the penitential processionals by the Flagellants during the Black Death, which had so devastating an impact on the European social order, arose from this famous account of how Pope Gregory stopped the plague in Rome.

As we will see, though, the nature of the Second Pandemic was very different from the first—and penitential processions, great masses in the cathedral, or any other activity that brought together large numbers of people, was the worst remedy for the Black Death that anyone could have devised.

* * *

A fairly clear picture of the dynamics of Justinian's Plague emerges from the sources. First, there is the pattern of spread. The disease followed the trade routes of Justinian's empire: from Ethiopia up to Egypt, and then throughout the Mediterranean basin. It did not penetrate northern Europe, or far into any countryside: this is in keeping with what we know of the habits of the black rat, readily transported on ships, but not an inland wanderer. This pattern immediately suggests rat-borne plague.

Our eyewitness accounts, too, do not mention contagion. They also omit that hallmark of contagion, the abandonment of the sick by their loved ones, that makes so many accounts of the Black Death such painful reading. Evagrius explicitly says that people who had lost too many of their family to want to live, and who deliberately kept company with the dying in order to contract the disease themselves failed to do so, as if the disease had a mind of its own.

As for the rate of plague, John of Ephesus tells us that it fell furiously upon a town, established itself, then moved along slowly. And Gregory of Tours makes clear that it took time for the disease to burst forth in the residential districts of port towns. This is the kind of dynamic we would expect from rat-borne plague.

And then there are the buboes, described at length, leaving us in no doubt about the symptoms of the epidemic. Justinian's Plague was classical rat- and rat-flea-borne bubonic plague, the kind of outbreak witnessed by the plague researchers in China and India during the Third Pandemic.

But where did the plague come from? Many scientists believe that the plague focus in Central Asia is probably the oldest one in existence, which they deduce from the great diversity of strains and variants circulating among the marmots and gerbils of those mountains and dry grasslands. These variants are biochemically distinct from the recent strains that, using the wide networks of twentieth-century shipping, established themselves across much of the earth. But why should the most closely related strains to those of Central Asia be found in East and Central Africa?[40]

The connection between Africa and Central Asia at first seems obscure. But though Central Asia and Central Africa seem worlds apart, animal species are more contiguous than one might think.

As Ken Gage explains, there is a vast gerbil belt that stretches all the way from Central Asia into Central Africa. There are various sorts of gerbils, but their habitats run and blend into each other. Many of these gerbils carry at least one species of *Xenopsylla*, the genus of fleas that causes plague in rats. Gage thinks that *Xenopsylla* species may very well have evolved on gerbils in the first place, and only later moved on to rats—specifically, the Nile grass rats of Egypt.[41]

Both grass rats and gerbils carry plague, and gerbils are thought to be a very ancient plague reservoir, perhaps, along with marmots, the oldest on earth. To think of plague spreading among these gerbils from Central Asia through the Middle East and down into Africa does not require a great stretch of the imagination. For gerbils, these continents are, in a sense, one long habitat, and the disease may have swept onward, like an electric current or a wave in the sea, right into Africa, where it smoldered among gerbils and Nile grass rats until conditions were right for a conflagration.[42]

What could have set the plague off? As we have seen, there appear to have been earlier plague outbreaks in Libya, Syria, and Palestine. In order to spill over into the colonies of domestic rats that were already inhabiting human settlements in the circum-Mediterranean, plague needs some sort of disruption of its natural patterns. Otherwise it endlessly cycles among the wild rodents that are its natural hosts: perhaps in infected fleas, perhaps among more resistant animals, perhaps in the cool, moist soil of animal burrows.

What this disruption means is that the wild rodents must, in some abrupt way, come into contact with domestic rodents, who in turn will contract the disease, die, and release their fleas to bite human beings. It is obvious from the shock and horror that greeted the Justinian Plague pandemic that this disease was a novelty to the inhabitants of Justinian's world. Earlier outbreaks had been long forgotten.[43]

Ken Gage insists that for a plague pandemic to be unleashed, a great many factors must be present. We do not know what all those factors are. But we can make a reasoned inference. Gage's research suggests that, under certain circumstances, cool, rainy conditions can precipitate outbreaks of prairie dog or ground squirrel plague in the American Southwest. Rainy conditions mean bumper crops of the kind of seeds and grasses these rodents eat: as a result, the prairie dogs or ground squirrels start producing more offspring. The prairie dogs cannot know that the next year, or the year after, things will change and their offspring will starve to death, or that the bursting population will destroy their local habitat, or that too many prairie dogs crowded together means that plague will transmit more easily, become more virulent, and wipe out their entire colony. Things are good for the nonce, so prairie dogs produce bigger litters than usual.

In the case of the gerbils and Nile grass rats in Justinian's day, Gage believes, unusual climatic events causing long-term cool, humid conditions could have had a similar effect. These animals could have outstripped the bounds of their habitats and crowded closer and closer to human dwellings. We know that these animals carry species of *Xenopsylla*, the most efficient plague vector, as well as other flea species that can also transmit plague. Crowded together with domestic rats, the fleas could easily enough have jumped species and infected those rats, who in turn unleashed plague upon their human associates, both locally, and among those who waited for grain shipments from Africa. The cloth shipped out of Egypt, too, could have carried infected rat fleas, as we know cloth and woven goods did during the Renaissance. We will see later how critical such shipments were for the maintenance of plague in Europe; quite probably infected fleas in cargo played a role in the Justinian pandemic as well.

There is quite a bit of evidence for a cooling trend in Justinian's day. Ice core sampling and tree ring data suggest that a giant volcanic explosion in the East Indies in around the year 535 had a dramatic impact on global weather. Particles of dust in the air caused cloudy, cool conditions for the year. The changing climate induced wild rodents—gerbils, grass rats, multimammate mice—to outstrip their habitat, sending hordes pressing up into the realm of domestic rats and people. Once human beings began dying of the plague, the disease could sometimes be spread by fleas alone: the cooler, moister climatic conditions could have helped infected fleas to survive much longer, nestled in the folds of finished Chinese silk or fine Egyptian linen, and make their way to the ports of Alexandria or Constantinople or Marseilles.

Once-smoldering wild rodent plague was thus unleashed upon Justinian's world, and the rest, unfortunately, is history.

* * *

The French scholars J.-N. Biraben and Jacques Le Goff have tracked every mention in the early medieval records of an outbreak of an epidemic disease either called *lues inguinaria* or *clades glandolaria* or presenting a description that seems to point to the "major symptoms" of plague.[44] They state that the first outbreak lasted from 541 to 544 C.E., and point out that this wave struck Byzantium much more sharply than it did the western reaches of the former Roman Empire: "Arriving probably by way of Genoa, Marseilles, and an undetermined Spanish port, the plague proceeded inland only as far as Clermont and Reims and subsided rather quickly."[45] Given what we have seen of the trade routes of Justinian's empire, this makes eminent sense.

Biraben and Le Goff list fourteen other waves, including the ones Evagrius records, and the Ligurian outbreak Paul the Deacon described two hundred years later. These successive waves lasted for about 150 years, and the effect was, Biraben and Le Goff tell us, "a demographic phenomenon of the first order." Its effect on the population of the circum-Mediterranean was "catastrophic."[46] What effect this dramatic decrease in the population of the empire had, especially in the cities, we can perhaps imagine: did the deaths of so many in the northeastern wing of the empire create a vacuum where Slavic tribes could rush in, to the Balkans and to Greece? Did it enable the Lombards to invade a greatly weakened Italy? The Berbers and Moors in their Atlas Mountain fastnesses were little affected by the plague that ravaged the North African coast—did the plague contribute to the revolt of the Berbers and the eventual destruction of the old Roman African civilization?

The Justinian Plague did not reach as far as the Black Death would reach, and must have killed far fewer people. But it has affected our world in ways we still do not fully understand. Some writers have suggested that the plague's depopulation of the Levant remains apparent to this day. Much former Byzantine farmland is still unsettled and uncultivated. The ruins of once thriving settlements, of towns and monasteries, can still be seen in the pasturelands of Syria. The Negev Desert, once home to many carefully nurtured monastic settlements, reverted to wasteland inhabited only by nomadic sheepherders until the mid-nineteenth century. The farmland of Africa dried up as irrigation systems fell into disrepair, and the Sahara began its inexorable advance into wheat fields that had fed an empire. The population itself did not recover for fifteen centuries, and even then the rural countryside remained sparsely settled; population growth took place largely in the major cities of the Middle East.[47]

As for the rise of Islam, how great a role did the relative weakness of the Byzantine Empire, after enduring many waves of plague, play in the ease with which that militant religion was able to sweep across North Africa and much of the Middle East? "Would it be unreasonable to suppose that the plague had something to do with the rather unexpected success of Arab revolts in the East and in North Africa?" ask Le Goff and Biraben.[48]

* * *

Historian Timothy Bratton points out that we can't say that "the plague, and only the plague, was responsible for all the ills of the post-Justinian Empire."[49] Much of the blame for those ills lies with the emperor himself. Justinian ought to have retrenched after the pandemic, and he did not. He had lost too much ground and too many men; the survivors of the plague struggled to pay the emperor's heavy taxes. The empire grew weaker, and the emperor's grasp on it began to slip. His attention began to drift away from the empire to theology; particularly after the death of Theodora in 548, he become obsessively concerned with the nature of the deity, and with settling disputes among the various Christian factions within the empire. He lost hope of Gaul and Britain; his success in Italy—he conquered the Ostrogoths definitively in 554—was temporary. A few years after his death, the invading Lombards swept into Italy, taking everything but Rome and surrounding lands, which they left to the Roman Church.

Western Europe crept out of the empire's shadow into the Middle Ages and its own history, leaving the classical world behind. It formed a new, syncretistic civilization, as tribe after tribe from the north settled down among the shards of old Roman Gaul. As Le Goff and Biraben put it:

> As for the West, there is one tempting hypothesis. It is a fact that the British Isles, northern Gaul, and Germania were, for the most part, spared by the plague. Could not this have been one of the reasons for the shift of power in Europe from the south to the north, from the Mediterranean to the North Sea? If we dared pursue this idea further—too far, no doubt—we might advance the hypothesis that the Justinian plague, having contributed to an explanation for Mohammed, can also explain Charlemagne.[50]

More recently, an American historian suggests that the course of

Western civilization owes something to the brutal interference of the plague in the Emperor Justinian's designs:

> That Western Europe was allowed to develop on its own, free from Imperial interference, to create many of the attitudes and institutions which are present in society to this day, may have been largely attributable to the foiling of Justinian's ambitions by the plague bacillus.[51]

# BLACK DEATH

**Pass not beneath, O Caravan, or pass**
**not singing. Have you heard**
**That silence where the birds are dead**
**yet something pipest like a bird?**

JAMES ELROY FLECKER

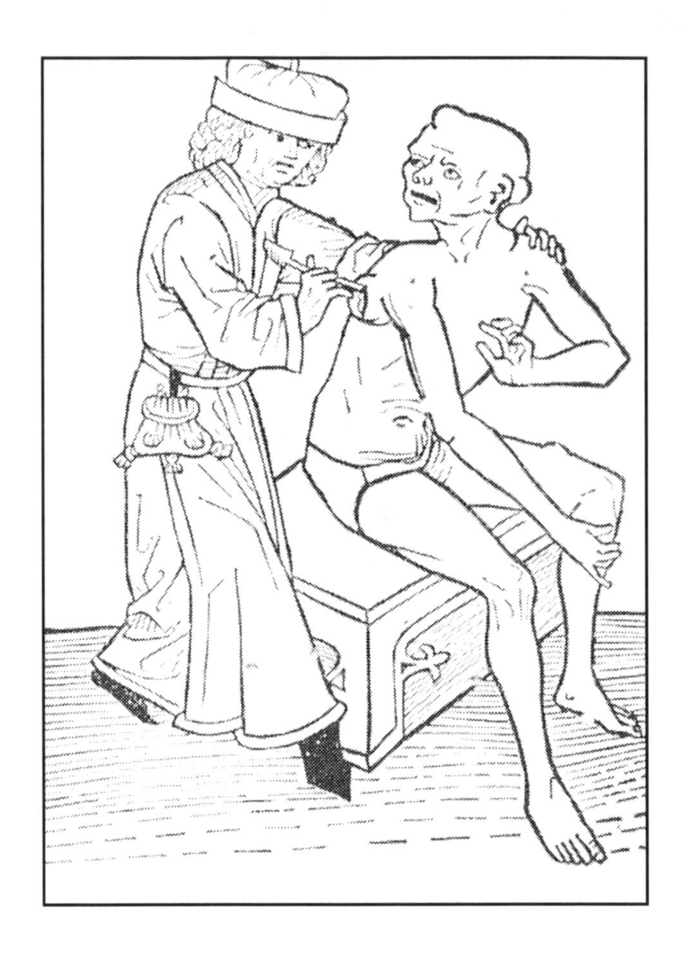

Medieval physician incising a bubo.

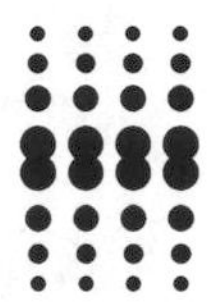

The story of the Black Death[1] begins long before 1347, the year that plague came once again to Europe.[2] As the First Pandemic may be laid, with some justice, at the Emperor Justinian's own door, so the Black Death may be charged to the cosmic account of another conqueror, one whose cold-eyed butchery made Justinian look a milksop by comparison. Chingis Khan, slaughterer of millions, knew nothing of the plague—he antedated the Black Death by 150 years. But he created the conditions whereby the Black Death could sweep unimpeded across Asia and Europe, reaching further and killing more than any Mongol warlord, more than Chingis himself, and more than Tamerlane.

Between the northern slopes of the T'ien Shan mountain range and the southern side of the Altai range in Central Asia are the high, treeless grasslands that have been the cradle of nightmare for the civilized world since the days of Herodotus. Wave upon wave of hard-eyed men, their skins weatherbeaten from the harsh suns and icy winds of the steppe country, their legs bowed from a life on horseback, fell upon the civilized world like wolves upon a sheepfold. The Scythians Herodotus described carried shields made of human skin and rode with the severed hands of their victims flapping along their saddles. Attila's Huns, squat, evil-smelling men, terrorized Europe and sacked Rome. Turkic-speaking peoples poured out of the East into Persia, the Levant, and even into Europe. Most of these tribes gradually settled alongside those they conquered, took up the plow and the appurtenances of civilization, and forgot their nomad ways. Softened by settlement, they themselves were prey for the next wave of destroyers.

The most terrible son of the steppe was a fatherless boy named Temujin, who was born clutching a clot of blood in his fist. His father was poisoned by a band of Tatars he did not recognize as old enemies; abandoned by his father's people, the boy, his mother, and several brothers had to survive by themselves in the bitter pastures of the steppe. Even then, Temujin would brook no opposition. Barely more than a boy, he murdered his half-brother, who had stolen a lark and a hare from him; even his mother, Hoelun, who reproached him, seems to have understood then what she had given birth to. Temujin grew up hard-eyed, calculating, and fearless; about him he gathered an army of disciplined riders the likes of which had never been seen before. These warriors might have frightened Attila himself. They never washed, believing that streams and lakes were sacred and not to be polluted by the human body; their stench was appalling. They were unbelievably hardy (they could endure all climates but the damp heat of India); their eyesight was keen, and their wiry strength and skill with the bow all served their purpose of terror and intimidation. They slaughtered anyone who opposed them down to the last child, leaving mountains of bones lying on the earth; they burned whole cities to the ground and laid waste cultivated lands to make more pasture. They cared for nothing but to extend their dominion to all corners of the earth, to everything that lay under the high blue carapace of their sky-god, Tengri.

Before Chingis Khan, as Temujin became known to history, sent his armies racing out across Asia and into Europe, the civilized world was divided into two realms united by a thin thread, the thread of the caravans along the Silk Road. (There was, also, the trade link through Persia that brought the silks of China into Byzantium.) This link could at any moment be severed—by bandits, by disease, by wild beasts; riding in a caravan through the deserts around Baghdad or the wastes of Central Asia was a dangerous business. The famed Silk Road—which actually followed a number of different tracks through various oases—was a fragile bridge between the Middle East (and, by extension, Europe) and the Far East.

The Mongols initially controlled only a small region of Central Asia, the area around the Onon and the Kerulen rivers, east of Lake Baikal. South of the Kerulen, by the shores of Lake Buyur, or Buir-Nor, and Lake Kolun, lived their ancestral enemies the Tatars; to the west were the Merkit, also hereditary enemies. The Merkit, who could call up some forty thousand horsemen,[3] kidnapped Temujin's wife, Borte, apparently out of revenge for Temujin's father's stealing of his mother, Hoelun. Temujin retrieved his wife and conquered the Merkit. It was his first victory. The Tatars, who had poisoned his father, were next. By 1206, all Mongolia lay at his feet. Chingis turned next to China, and conquered vast swaths of it, though China would not fall completely under Mongol rule until after his death. His first impulse was to slaughter the millions of Chinese peasants now under his dominion—he did not like agriculture and thought the land, which could be turned into pasture, wasted by the peasants and their patient farming. He was only turned aside by his wise Khitan advisor, Ye-lu Ch'u-ts'ai, who pointed out that peasants can be taxed, and that Chingis had better not slaughter these cash cows, so to speak. Apparently Ye-lu Ch'u-ts'ai knew better than to raise humanitarian concerns with the Mighty Khan.[4]

Chingis's forces then conquered the high steppe kingdom of the Kara-Khitai, a Turkic-Muslim population, and the Khwarizmian Empire: these areas encompass much of present-day Kazakhstan, Uzbekistan, Turkmenistan, and Iran. After these conquests, Temujin's forces moved up through the Caucasus and into Russia, eventually devastating much of Eastern Europe. The appalling massacres of any city that resisted him terrified Europe. "His destruction of eastern Iran exceeds in horror anything attributed by Europe to Attila," says the famed French scholar René Grousset.[5]

Chingis Khan (the Mighty Ruler, the name the conquering

Temujin gave himself in 1206), had more than the slender web of caravan trade to unite his empire. His men thundered through countless kingdoms; unafraid of bandits, they rode at their ease from China to Europe. Their supply lines were strong and their communication network astounding: the hardy Mongol riders linked the empire in a sort of large-scale pony express. "They created a territorially vast human web that linked the Mongol headquarters at Karakorum [in Mongolia] with Kazan and Astrakhan on the Volga, with Kaffa in the Crimea, with Khanbaliq in China and with innumerable other caravanserais in between," as William H. McNeill puts it in *Plagues and Peoples*, his landmark study of the effects of disease on history.[6] Even after the division of Chingis's empire into four separate domains, the worlds of East and West were united as they never had been before. Grousset puts it this way: "His *yasaq* established throughout Mongolia and Turkestan a "Pax Jenghiz-khana," no doubt a terrible one in his own day, but one which under his successors became milder and rendered possible the achievements of the great travelers of the fourteenth century."[7]

It was this strong cord between East and West, the security of the caravans, the passage of highly mobile armies, that allowed plague to leave its ancestral home in the high steppes of Mongolia. But this was no ordinary plague. The Justinian Plague and its succeeding waves, which still washed over Europe and the Middle East in sporadic fashion, had killed millions over the centuries. Even so, Justinian's Plague was predominately rat-borne plague, and relatively sluggish. But the plague unleashed unwittingly by the Mongols upon the whole known world was a plague of a different order.

Chingis died in 1227, at the age of sixty. After his death, his empire was divided into four *ulus* ruled by his four sons. They divided the Mongolian steppe land and its pasturage among them, though not the settled lands of China, Turkestan, and Persia, which at first they held in common.[8] Eventually, however, even these lands were divided under the dominion of different Mongol lords, and became separate domains. Of these realms, it is that of Chagatai, Chingis's second son, that most concerns us. Of all Chingis's sons, he was the least ambitious: it was his role to maintain the *yasaq*, his father's code of law. Chagatai held the lands that corresponded to the old Kara-Khitai region south of Lake Balkhash, which roughly corresponds to part of southern Kazakhstan, Tajikistan, Kyrgyzstan including the T'ien Shan mountains and the deep blue Lake Issyk-Kul, and part of eastern Turkestan (western China today). Chagatai's realm was also, as it is today, an inveterate focus of plague: the plague germ has circulated for millennia among the susliks, gerbils, and marmots of the region.

It was in the lands of Chagatai that, so far as we know, the Black Death claimed its first victims. The hunters of the Chagatai khanate were both in their location and mode of life likely to be exposed to plague. By the fourteenth century, Mongols in the other khanates had abandoned the lives of hunters and herders for the cities they had conquered. Kublai Khan, Chingis's grandson, eventually conquered all of China and settled in Beijing, where he built gardens and pleasure palaces. Hulagu conquered the Persians, and his soldiers learned their civilized ways. On the Volga, the Golden Horde ruled all Russia with a mailed fist for two centuries—but Western travelers from the fourteenth century gaped at the splendor of their capital, the Great Serai, with its wide streets, golden domes, and learned men.

Only the men of Chagatai's dominion kept to the old life of the nomadic pastoralist, hunting, roaming, moving their flocks from winter to summer ranges. They never abandoned their yurts for the cities. When the khans needed money, they sacked their own cities, Samarkand and Bukhara.

Like Chingis, too, they worshipped Tengri, the Eternal Blue Heaven. But, as Chingis had stipulated in his legal code, they were remarkably tolerant of other faiths.[9] This code, a bewildering farrago of prohibitions and punishments (adultery, desertion, theft, and a third bankruptcy for merchants were punishable by death), prohibited the ritual slaughter of animals in the Muslim fashion, and washing or urinating in streams.[10] It also forbade the taxing of the clergy of any faith. In Chagatai's realm both Buddhists and Nestorian Christians flourished.

In a Nestorian Christian cemetery near the blue mountain Lake Issyk-Kul we see the earliest traces of the Black Death. In the nineteenth century, Russian archaeologist Daniel Abramovich Chwolson, an expert in deciphering and interpreting ancient grave inscriptions, discovered three plague graves in this cemetery, all bearing the date 1650—or, in modern chronology, 1339. One stone read, "This is the grave of Kutluk—he died of plague with his wife Mangu-kelka."[11]

There is other evidence that something unusual happened that year in the Issyk-Kul Nestorian community. Six hundred fifty Nestorians had been buried in that graveyard over a 150-year period. But in this one year (1338–1339) there were over one hundred graves—a death rate elevated by a factor of ten.[12]

As microbiologist Mark Wheelis of the University of California at Davis points out, we cannot be absolutely certain that these Nestorian Christians died of true plague. But the area, an endemic plague focus, the timing (not long before the outbreak of the Black Death in Europe), the statement on the gravestones, and the massive increase in deaths point to the same conclusion. The Black Death had fired its first shots, and these Nestorian Christians, living quiet lives deep in Mongol country, were among its early targets.

Finding these early graves gives us a strong indication of where the Black Death began its depredations. But it does not tell us how the plague became in this instance a human disease. For each outbreak there has to be a portal through which plague enters the human species. In the Justinian Plague a likely trail is laid out for us: gerbils to Nile grass rats, grass rats to domestic black rats, black rats to people. Here, the culprit species is a much more ancient reservoir, and much more dangerous: the marmots—tarabagans and bobaks—of Central Asia. Marmot strains, according to Domaradskij and other Russian scientists, are the most virulent of all known strains of plague. The plague germ has been waging a long battle with the marmots of Central Asia; as these marmots, over great spans of time, grow more resistant, the plague germ that circulates among them grows ever more virulent. Recent work by the Russian plague experts V. V. Suntsov and N. I. Suntsova of the Severtsov Institute of Ecology and Evolution in Moscow also suggests that the most ancient plague strains are, indeed, found in the marmots of Central Asia.[13]

As for the connection between marmots and Mongols, this, too, is ancient. Marmots are ubiquitous throughout the high grasslands, where, like giant prairie dogs, they live in vast colonies of underground burrows. For centuries, if not millennia, the inhabitants of the Central Asian steppe must have hunted them for food and fur. According to Mongolia scholar J. Batbold, between the years 1906–1994 over 100 *million* marmot skins were prepared in Mongolia.

It is well known that Mongols and Manchurians today do not hunt slow-moving marmots, as they believe that sluggish animals are probably diseased. Marmots also must be shot and not trapped, so the hunter is less likely to bring home a plague-stricken animal. Despite these precautions, marmot plague still occasionally causes explosive epidemics.

Perhaps similar rules were in place in the fourteenth century; we have no way of knowing. But in any case it requires little stretch of the imagination to suppose that some Mongol or Nestorian hunter, sometime in the fourteenth century, caught and skinned a plague-infected marmot, and released a whirlwind.[14]

This marmot plague spread from person to person in a long

chain of disease and death. The Black Death was, from its onset, more lethal, more terrifying, and much more contagious than any plague outbreak borne by rats. It was a fitting legacy of Chingis Khan, one of the greatest mass murderers in history.

* * *

Contemporary writers most in a position to know attributed a Mongolian origin to the Black Death. Historian Michael Dols, in his history of plague in the Middle East, quotes a fourteenth-century Muslim writer, Ibn al-Wardi, who witnessed the initial outbreak in Aleppo, Syria, and later died of it (in 1349).[15] According to Ibn al-Wardi, the Black Death came from "The land of darkness." By this Ibn al-Wardi meant inner Asia or Mongolia, where, he says, the plague raged for fifteen years before spreading out-ward to the rest of the world.[16] In his account, the plague moved south and east from Mongolia to China and India, and westward "to the land of the Khitai"—part of Chagatai's domain.[19] Then it moved "to the land of the Uzbeks"—though some scholars feel that this actually refers to the realm of the Ozbeks of the Golden Horde, otherwise known as the Kipchaks. This is just north of Chagatai's khanate. From there, Dols states, the plague moved on to "Transoxiana, Persia, and finally to Crimea and the Mediterranean world."[20] Another important source, the Egyptian Al-Mazriqi (1364–1442), states that the disease began "a six-months' journey from Tabriz" (Persia) in the lands of the Khitai and Mongols; as Dols puts it, these people comprised "more than three hundred tribes [who] all perished without apparent reason in their summer and winter encampments, in the course of pasturing their flocks and during their seasonal migration."[17] Countless of the khan's soldiers died, Al-Mazriqi notes, and the land was left empty all the way to Korea; the khan himself, and six of his children, perished as well. All this took place amid a series of terrifying environmental catastrophes—torrential rainfalls out of season, earthquakes, the deaths of animals.[18] Al-Mazriqi, a famous historian of the Mameluke Empire, was not yet born when the Black Death came to Cairo in 1348, though he did describe later visitations that he witnessed.

Dols cites other Arab medieval chroniclers who also claimed that plague began in the land of the Khitai: Ibn al-Khatimah and Ibn al-Khatib, both of Moorish Spain. Ibn al-Khatimah, who had it from Christian merchants that the Black Death began in Khitai, thought they were speaking about China; Ibn al-Khatib wrote that the outbreak first began in Khitai and in the Indus Valley. Ibn al-Khatib, a learned, sensitive, and keen observer, readily perceived that the plague was highly contagious, spreading rapidly from man to man. The prophet Muhammad had taught that illnesses, particularly plagues, were a gift from God, and belief in contagion, therefore, became heresy in the Muslim world. Ibn al-Khatib's insistence on the obvious fact of plague's contagion flouted religious authority, and angered the Muslim leadership of his time. Nonetheless, Ibn al-Khatib recognized that plague spreads most when the lungs are infected.[21] In due course, the heretical Ibn al-Khatib paid for his conviction with his life.[22]

The route of the plague must have passed from Chatagai's khanate east to China and west along the one trade route left after the shattering of Chingis's empire into four khanates. Mongolian Persia, the khanate of Hulagu, had collapsed by the thirteenth century; it was ridden by internecine struggles among the contenders to Hulagu's domain. But trade through the northern route was as steady as ever; enough was left of the "Pax Jenghiz-khana" to ensure the safe passage of a steady flow of people and caravans from the Far East to the Black Sea. "By the fourteenth century," Dols tells us, "the northern transcontinental route ran from the Genoese and Venetian counting-houses of the Crimea to Peking. The principal stages were Sarai on the Lower Volga, Otrar, Talas, and Balasagun, west of Lake Issyk-sKul," where the earliest known plague cemetery was found.[23]

Furthermore, there were armies on the move, armies that were highly mobile and could swiftly carry infection. Plague broke out in Azerbaijan in 1347 while one contender for Hulagu's domain,[24] Malik Ashraf, laid siege to Tabriz. Later Ashraf's army brought the plague to Baghdad as well.[25] It seems that plague diffused south from Chagatai's domain through the Caucasus and into Persia, along the trade routes and the paths of armies.

It diffused through the Golden Horde as well—the Kipchak khanate that bordered Chagatai's domain on the north. In 1345–1346, plague broke out in the Great Sarai, the Golden Horde's capital on the Volga, and in other cities in the khanate. It depopulated the Golden Horde and spread into the Crimea and Byzantium, though it did not reach Moscow until much later (1351) and by quite another route.

The city of Kaffa—present-day Feodosiya on the Black Sea—may have been the portal through which plague spilled into Europe. It was then (1345) a Genoese stronghold; in 1266, the government of the Golden Horde had leased the site—which had once been a Greek city until Attila and his Huns destroyed it in the fourth century C.E.—to traders from Genoa; by the fourteenth century, a thriving Genoese colony and fort had sprung up. Venetian traders occupied the nearby trading city Tana. But relations between the Italian traders and the Golden Horde were tense and often hostile; in 1343, after a brawl between Muslims and Italians in Tana, Khan Janibeg ordered all the Genoese and Venetians to leave. Then he laid siege to Kaffa. The blockade of Kaffa could not be complete, as the Genoese had the Black Sea at their backs. The siege failed; but two years later Janibeg's men returned—and the plague broke out in the Mongol ranks outside the city gates.

No eyewitness record of that outbreak and its direct consequences is left to us. But Gabriele de' Mussis, a lawyer of Piancenza, wrote a famous account of the fate of Mongols and Genoese alike:

> O God! See how the heathen Tartar races,[26] pouring together from all sides, suddenly invested the city of Caffa and besieged the trapped Christians there for almost three years. There, hemmed in by an immense army, they could hardly draw breath, although food could be shipped in, which offered them some hope. But behold, the whole army was affected by a disease which overran the Tartars and killed thousands upon thousands every day. It was as though arrows were raining down from heaven to strike and crush the Tartars' arrogance. All medical advice and attention was useless; the Tartars died as soon as the signs of disease appeared on their bodies: swellings in the armpit or groin caused by coagulating humours, followed by a putrid fever.
>
> The dying Tartars, stunned and stupefied by the immensity of the disaster brought about by the disease, and realizing that they had no hope of escape, lost interest in the siege. But they ordered corpses to be placed in catapults and lobbed into the city in the hope that the intolerable stench would kill everyone inside. What seemed like mountains of dead were thrown into the city, and the Christians could not hide or flee or escape from them, although they dumped as many of the bodies as they could in the sea. And soon the rotting corpses tainted the air and poisoned the water supply, and the stench was so overwhelming that hardly one in several thousand was in a position to flee the remains of the Tartar army. Moreover one infected man could carry the poison to others, and infect people and places with the disease by look alone.[27]

This has often been taken as an early incidence of biological warfare, and it may be so in fact, though we have no way to know whether these famous catapults with their noisome missiles were actually ever fired. The only reference to this incident is the account of de' Mussis, and it has been shown that the Piancenza lawyer was firmly ensconced in his hometown throughout the events he so vividly describes.[28] Could the plague really have been spread in this way? Igor Domaradskij says the threat from dead bodies is "theoretical only"—but these were not merely dead bodies, they were rotting, oozing bodies hurled from a great distance and smashed against the stone walls and streets of a medieval city.[29] People within could have been directly exposed to infected fluids in that manner.

There is another account of the siege at Kaffa, by a French abbot named Gilles Li Muisis, which does not even mention those famous plague catapults:

> I heard that in the previous year, 1347, an innumerable horde of Tartars laid siege to a very strong city inhabited by Christians. The calamitous disease befell the Tartar army, and the mortality was so great and widespread that scarcely one in twenty of them remained alive. After discussing it among themselves, they came to the decision that such a great mortality was caused by the vengeance of God, and they resolved to enter the city which they were besieging and asked to be made Christians. Accordingly the most powerful of the survivors entered the city, but they found few men there for all the others had died. And when they saw that the mortality had broken out among the Christians as well as among themselves, because of the unhealthy air, they decided to keep to their own religion.[30]

As both these accounts are based on hearsay, there is no reason on the face of it to prefer one account to the other.

How plague entered Kaffa we cannot know, but there is little doubt about what happened next. Though the Mongol siege was lifted at some point after the Black Death devastated the Mongol army, there never was a blockade on the sea; among those who escaped from Kaffa were, as de' Mussis puts it, "a few sailors who had been infected with the poisonous disease." Everywhere they landed, de' Mussis writes, "it was as if they brought evil spirits with them: every city, every settlement, every place was poisoned by the contagious pestilence. . . . And when one person had contracted the illness, he poisoned his whole family even as he fell and died."[31]

Though Justinian's Plague was also spread by ship, the pattern of this pandemic is clearly very different. There was never any mention of contagion in the accounts of the earlier plague. But this theme of deadly contagion—of people being poisonous to their friends and family—is one that occurs, again and again, in accounts of the Black Death.

By 1348 the plague reached Italy. De' Mussis states that "scarcely one out of seven of the Genoese survived," while 70 percent of Venetians died, including twenty out of twenty-four "excellent physicians"—another index of contagiousness.[32] (Procopius, during the Constantinople outbreak of 542, made clear that those attending victims of the plague did not contract the disease.)

De' Mussis insists that goods could also transmit the illness. Four soldiers left a Genoese division camped near the city in search of plunder; they traveled to Rivarolo on the coast, which was left entirely deserted. They broke into an empty house and stole a fleece they found upon a bed. The next night, having rejoined their army, they bedded down under the fleece. The next morning their fellows woke to find them dead.[33]

More believable is de' Mussis's account of how the plague came to Piancenza, which he would have known firsthand.[34] One of the Genoese, already stricken, came to Piancenza and sought shelter with a friend, Fulco della Croce. The Genoese went to bed and died immediately; Fulco, his entire family, and many of their neighbors died as well. The disease spread throughout the city; every day "one sees the Cross and the Host being carried throughout the city and countless dead being buried."[35]

The way de' Mussis describes this outbreak recalls the great plague fighter Robert Pollitzer's distinction between zootic and demic plague. Zootic plague, borne by rats and rat fleas, typically strikes one person in a household and does not spread. Demic plague wipes out entire households.[36] Whether it is person-to-person flea-borne plague or pneumonic plague, or both, we can't be certain. As Wu Lien-teh points out in his 1926 *A Treatise on Pneumonic Plague*, cases of bubonic and septicemic plague occur alongside pneumonic plague during pneumonic epidemics. These are caused, he says, "either by direct contact with pneumonic-plague victims, or contaminated objects, or through human parasites."[37] As to symptoms, de' Mussis is deliberately specific, "so that the conditions, causes, and symptoms of the pestilential disease should be made plain to all." He describes

> four savage blows to the flesh. First, out of the blue, a kind of chilly stiffness troubled their bodies. They felt a tingling sensation, as if they were being pricked by the points of arrows. The next stage was a fearsome attack which took the form of an extremely hard, solid boil. In some people this developed under the armpit and in others in the groin between the scrotum and the body. As it grew more solid, its burning heat caused the patients to fall into an acute and putrid fever, with severe headaches. . . . In some cases it gave rise to an intolerable stench.[38] In others it brought vomiting of blood. . . . Some people lay in a drunken stupor and could not be roused. . . . There was no known remedy for the vomiting of blood. . . . But from the fever it was sometimes possible to make a recovery."[39]

One way or another the plague must have first made its way to Europe from the Crimea. But the timing in de' Mussis's story makes little sense.[40] Plague struck the Mongols sometime in 1346; it did not reach Genoa until 1348; actually, the first Italy ports struck by plague were in Sicily, in 1347. In the spring of that same year, long before it reached Genoa, plague struck Constantinople; we have the vivid description of its course by Emperor John Cantacuzene himself, whose own son, Andronicus, died of it. "[Some] had a very violent fever at first, the disease in such cases attacking the head; they suffered from speechlessness and insensibility to all happenings and then appeared as if sunken into a deep sleep. . . .

"Sputum suffused with blood was brought up and disgusting and stinking breath from within. . . . Great abcesses were formed on the legs or the arms, from which, when cut, a large quantity of foul-smelling pus flowed.[41] The epidemic which then [1347] ranged in northern Scythia traversed almost the entire sea-coasts, whence it was carried over the world. For it invaded not only Pontus, Thrace, and Macedonia, but Greece, Italy, the Islands, Egypt, Lybia, Judea, Syria, and almost the entire universe."[42]

Ships passing through from the Black Sea to the Mediterranean would necessarily have passed through the Golden Horn of Constantinople. Did it take six months, at least, for plague to move on these death ships from Kaffa to Constantinople? Another year to reach Genoa? Considering that it took Christopher Columbus two months to cross the Atlantic, it is impossible to imagine that plague came directly to Genoa from Kaffa on a voyage of a year and a half. Furthermore, we see that the plague in Genoa, Piancenza, and elsewhere was highly contagious. A trip of even several months' duration under such conditions is also unimaginable. The sailors crowded together in the hold would have fallen ill and died long before the ship could put into port.[43]

Furthermore, a source other than de' Mussis, an anonymous Flemish scribe, states that in January of 1348 "three galleys put in at Genoa, driven by a fierce wind from the East, horribly infected and laden with a variety of spices and other valuable goods."[44] Desperately ill soldiers fleeing a besieged city would not come home laden with spices and goods. These were trading galleys; the mention of spices suggests the goods came from trade with the Far East, which must have passed through Mongol country.

This is a more likely scenario than de' Mussis's. Plague must have come into Italy not once, but many times, from many points along the way to the Mediterranean from the Crimea. Though Kaffa may have been the initial portal for plague's entry into Europe, it is likely that the plague was spread in stages along the way to the Mediterranean and to Europe.[45]

* * *

The appalling dimension of the disaster that would now strike Europe is difficult for us to grasp. In our modern experience, only the mass slaughter of the innocent—the killing grounds of Auschwitz and the Gulag, the mounds of skulls left by Pol Pot and his Khmer Rouge, the butchery of innumerable Rwandans—gives us any sense of such a disaster, of a world where the fabric of ordinary life is rent completely, where the unregarded dead are stacked in their multitudes and left to rot. The great Florentine poet Petrarch, in a letter to his dear friend Louis Heyligen, writes, "There was a crowd of us, now we are almost alone. We should make new friends, but how, when the human race is almost wiped out; and why, when it looks to me as if the end of the world is at hand? Why pretend? We are alone indeed. . . . How transient and arrogant an animal is man! How shallow the foundations on which he rears his towers! You see how our great band of friends has dwindled. Look, even as we speak we too are slipping away, vanishing like shadows. One minute someone hears that another has gone, the next he is following in his footsteps."[46]

Petrarch and his friend Giovanni Boccaccio were the most cultivated of men, distinguished by their breadth of intellect and largeness of spirit from most of their contemporaries.[47] They remain eloquent witnesses to a catastrophe that most people suffered uncomprehendingly, terrified into mass hysteria, bloody self-flagellation, helpless penitence, or silence.

Michael of Piazza, a Franciscan friar, describes the first appearance of the Black Death in Europe:

> At the beginning of October, in the year of the incarnation of the Son of God 1347, twelve Genoese galleys were fleeing from the vengeance which our Lord was taking on account of their nefarious deeds and entered the harbor of Messina. In their bones they bore so virulent a disease that anyone who only spoke to them was seized by a mortal illness and in no manner could evade death. The infection spread to everyone who had any intercourse with the diseased. Those infected felt themselves penetrated by a pain throughout their whole bodies, and, so to say, undermined. Then there developed on their thighs or under their upper arms a boil about the size of a lentil which the people call "burn boil" (antrachi). This infected the whole body, and penetrated it so that the patient violently vomited blood. This vomiting of blood continued without intermission for three days, there being no means of healing it, and then the patient expired. But not only all those who had intercourse with them died, but also those who had touched or used any of their things.[48]

It began in the city of Messina in Sicily, borne by several Genoese galleys. The people in Messina, when they understood that the sailors carried death in their ships, promptly drove them off. But it was too late. The deaths in Messina began to mount; many ecclesiastics, fearing for their own lives, refused to enter the houses of the sick. Whoever heard the confession of the dying died as well; some, says Michael, were struck down in the same rooms as the dying—though this hardly seems likely; even septicemic plague takes longer than several hours to kill.

"Soon the corpses were lying forsaken in the houses," Michael continues. "No ecclesiastic, no son, no father and no relation dared enter, but they paid hired servants to bury the dead. . . . When the catastrophe had reached its climax the Messinians resolved to emigrate. One portion of them settled in the vineyards and fields, but a larger portion sought refuge in the town of Catania, trusting that the holy virgin Agatha of Catania would deliver them from their evil."

But the Catanians, seeing how plague began to spread in their city, drove the Messinians away, with a heartless "Be gone, you are from Messina!" The now infected Catanians held massive processionals bearing the bones of their own holy virgin Agatha, giving the infection, not incidentally, an even better chance to spread. "The town of Catania lost all its inhabitants, so that it ultimately sank into complete oblivion." The Messinians, meanwhile, still seeking refuge, spread all over the island of Sicily, and into Syracuse. The town of Trapani was, like Catania, left completely desolate.

The friar was exacting in his description of physical symptoms:

> Here not only the "burn blisters" appeared, but there developed in different parts of the body gland boils in some on the sexual organs, in others on the thighs, in others on the arms, and in others on the neck. At first these were of the size of a hazel-nut and developed accompanied by violent shivering fits, which soon rendered those attacked so weak that they could no longer stand upright, but were forced to lie in their beds consumed by violent fever and overcome by great tribulation. Soon the boils grew to the size of a walnut, then to that of a hen's egg or a goose's egg, and they were exceedingly painful, and irritated the body, causing it to vomit blood by vitiating the juices.

These descriptions tell us several things: that both bubonic and pneumonic forms were present in Sicily; that the disease was extraordinarily contagious and spread very rapidly. We do not appear to have the exact timetable of spread. But from this and other accounts the inescapable impression is of a highly contagious disease that did not need to wait, as the Justinian Plague did, for infected rats to desert a ship and infect other local rats before the disease blazes out in the human population.[49] One of the sharpest differences between Justinian's Plague and the Black Death was that the latter reached much deeper inland, while the earlier plague was largely confined to the ports.

The physician Simon de Covino, a Parisian doctor who seems to have been in Montpellier when that Mediterranean city was struck with the plague, identifies the plague as *pestis inguinaria*, the bubonic plague of the East.[50] The disease began as a burning pain at the site of the future bubo, and developed into a fever that attacked the internal organs of the body, especially the heart and lungs. "No climate," says de Covino, "appeared to have any effect on the strange malady. It appeared to be stayed by neither heat nor cold. High and healthy situations were as much subject to it as damp and low places. It spread during the colder season of winter as rapidly as in the heat of summer."[51]

De Covino also makes clear that, unlike Justinian's Plague, the Black Death spread by a violent contagion. "It has been proved that when it once entered a house scarcely one of those who dwelt in it escaped. . . . [One sick person] "would infect the whole world. . . . It happened also that priests, those sacred physicians of souls, were seized by the plague whilst administering spiritual aid; and often by a single touch, or a single breath of the plague-stricken, they perished even before the sick person they had come to assist."[52]

The pattern, or footprint, of the Black Death, therefore, is quite distinct from the earlier plague pandemic. Justinian's Plague was confined to the cities directly reached by the commercial network of Justinian's empire. The Black Death, being directly contagious beyond a doubt, reached much further and struck much harder.

Though Justinian's Plague laid waste much of the circum-Mediterranean area, the Black Death stretched across the known world, penetrating deep into cities and countryside far from the sea. It is this geographical range, in part, that has made plague skeptics suspicious, and with reason, because it is really quite difficult to imagine plague-infected rats, who are not nomadic even when they are healthy, trundling across the vast continents of Europe and of Asia. The Black Death could spread as fast as people could travel, because it was borne by human beings.

* * *

The plague reached Sicily in October of 1347. Three months later, by early January 1348, carried once again by ships, it struck Genoa on the west coast of Italy and Venice in the east. The disease next appeared along the Neapolitan coast of Italy, carried by two cargo ships to Pisa.[53] Pisa served as the funnel that spilled the plague into Tuscany and throughout the Italian interior, where it quickly invaded both cities and countryside, spreading across to Florence and down into Rome.

Florence, at that time one of the major cities of Europe, with close to 100,000 citizens, was a sophisticated polity, though racked by political upheavals.[54] The most eloquent witness to the Florentine plague was the poet and litterateur Giovanni Boccaccio, close friend of Petrarch's, and a man whose eyewitness account of the Black Death is analogous to Procopius's in its significance for the historical record. Boccaccio was that rarity in the medieval world, a man who chose the life of the mind and yet eschewed the Church. He was raised to enter commerce, but despised it; his father next induced him to study law, to no happier effect. Eventually, his father gave up and allowed him to pursue the study of poetry and literature; it was a difficult existence and the younger Boccaccio battled poverty all his life.

Boccaccio is chiefly remembered today for his *Decameron*, a series of tales set during the plague in Florence. Ten young people, seven women and three men, decide to retire from the plague-struck city and spend their days pleasantly in the country retreat of one young woman, where they ate delicately, drank wine, and passed the time by telling stories. It is the famous Prologue to the *Decameron* that describes the actual circumstances of the plague in Florence as Boccaccio witnessed them. That the plague in Florence was devastating goes without saying; that it was contagious as well there can be no doubt:

> Let me say . . . that thirteen hundred and forty-eight years had already passed after the fruitful Incarnation of the Son of God when into the distinguished city of Florence, more noble than any other Italian city, there came the deadly pestilence. It started in the East, either because of the influence of heavenly bodies or because of God's just wrath as a punishment to mortals for our wicked deeds, and it killed an infinite number of people. Without pause it spread from one place and it stretched its miserable length over the West. And against this pestilence no human wisdom or foresight was of any avail; quantities of filth were removed from the city by officials charged with this task; the entry of any sick person into the city was prohibited, and many directives were issued concerning the maintenance of good health. . . . Nor were the humble supplications, rendered not once but many times to God by pious people, through public processions or by other means, efficacious; for almost at the beginning of springtime of the year in question the plague began to show its sorrowful effects in an extraordinary manner. It did not act as it had done in the East, where bleeding from the nose was a manifest sign of inevitable death, but it began in both men and women with certain swellings either in the groin or under the armpits, some of which grew to the size of a normal apple and others to the size of an egg (more or less), and the people called them *gavocciolo*.[55] And from the two parts of the body already

mentioned, within a brief space of time, the said deadly *gavoccioli* began to spread indiscriminately over every part of the body; and after this, the symptoms of the illness changed to black and livid spots appearing on the arms and thighs, and on every part of the body, some large ones and sometimes many little ones scattered all around. And just as the *gavoccioli* were originally, and still are, a very certain indication of impending death, in like manner these spots came to mean the same thing for whoever had them. . . .

This pestilence was so powerful that it was communicated to the healthy by contact with the sick, the way a fire close to dry or oily things will set them aflame. And the evil of the plague went even further: not only did talking to or being around the sick bring infection and a common death, but also touching the clothes of the sick or anything touched or used by them seemed to communicate this very disease to the person involved. . . . Let me say, then, that the power of the plague described here was of such virulence in spreading from one person to another that not only did it pass from one man to the next, but what's more, it was often transmitted from the garments of a sick or dead man to animals that not only became contaminated by the disease, but also died within a brief period of time. . . .

Oh, how many great palaces, beautiful homes, and noble dwellings, once filled with families, gentlemen, and ladies, were now emptied, down to the last servant! How many notable families, vast domains, and famous fortunes remained without legitimate heir! How many valiant men, beautiful women, and charming young men, who might have been pronounced very healthy by Galen, Hippocrates, and Aesculapius (not to mention lesser physicians), dined in the morning with their relatives, companions, and friends and then in the evening took supper with their ancestors in the other world![56]

Two things, in this context, are noteworthy about this account:

first, Boccaccio does not mention blood-spitting—he even denies that people bled from the nose as they did in the East—and second, he insists upon the disease's extreme contagion. Taking these two aspects together, the transmission pattern Boccaccio describes suggests human-to-human flea-borne transmission, probably by way of the so-called human flea, *Pulex irritans*. That fleas were common in medieval Florence there can be little doubt: a description of an Italian pesthouse describes the "clouds of fleas" that infested moribund patients. Furthermore, the pattern of the disease itself sounds as if numerous flea bites producing many sites of infection and multiple buboes, developing into rapid septicemia characterized by the black and livid blotches, were common in this outbreak. While infection by means of *Pulex irritans* remains controversial in the United States, a Russian-language book, *Fleas As Carriers of Germs of Human and Animal Diseases*, identifies *Pulex irritans* as an "active carrier of plague."[57] French scientists Georges Blanc and Marcel Baltazard, writing several important studies from the 1940s to the 1960s based on their extensive research in Iranian Kurdistan and in Morocco, insist that *Pulex* was a dominant force in spreading interhuman infection, both in those countries and in medieval Europe.[58] Also, the Kazakh scientists Suleimenov and Atshabar insist that transmission through *unblocked* fleas such as *Pulex* is much more significant for the spread of human plague than through the blocked fleas like the Oriental rat flea classically supposed to be the main vector of plague. Furthermore, unblocked fleas such as *Pulex* are likely to transmit disease only from cases with a high level of plague germs in the blood—in other words, from people with septicemic plague. The Black Death was from its inception an extraordinarily virulent form of plague. It is quite likely that such virulent plague could produce septicemia, and therefore be passed, via *Pulex*, directly from person to person.

Kenneth Gage of the CDC has come, in recent years, to believe that *irritans* played a far more important role in the Black Death than is commonly thought, as the evidence for transmission between people is so clear from records such as Boccaccio's. He has his own indirect evidence as well. When Gage and his team investigated small plague outbreaks at high elevations in the Andes—in Ecuador and Peru—the huts they visited were "absolutely loaded with *irritans.*"

"We were biased from our grad school education on plague," says Gage, "that it had to be blocked fleas or nothing. But when we dragged woolen blankets out of the huts, *irritans* were just popping off everywhere, begging to be noticed.

"And transmission via *irritans* as well as pneumonic plague—either one could favor increased virulence."

Certainly Boccaccio's insistence on the rapid fatality of the illness seems to suggest septicemic plague, especially the famous passage about dining with one's friends at lunch and with one's ancestors at dinner—even though it is unlikely that the most virulent plague could kill so fast, and so we must grant him some measure of poetic license.[59]

But sources other than Boccaccio make clear that pneumonic plague was also still present as the Black Death penetrated Europe. At the same time that plague reached Genoa (January 1348), it also came to Avignon, the city in southern France that was, for two centuries, the seat of the papacy. Avignon lies northwest of Marseilles, which was the point of entry into France for the Black Death, in November 1347,[60] as it had been, in the past, for Justinian's Plague, and would be again in 1720, in the last great outbreak of plague in Western Europe. According to the French historian Biraben, the path plague took from Marseilles into the country seems to have been through Aix-en-Provence (reached before Christmas 1347), which lies some thirty kilometers from Marseilles. From Aix, the path led to Arles (74 kilometers) and Avignon (75 kilometers); from Arles, the plague reached Narbonne in the southwest (172 kilometers) in March; Carcassonne, Toulouse, and Perpignan by June. By the end of June, the disease had reached Bordeaux and the Atlantic Ocean in the west, and was moving southward over the Pyrenees into Spain.[61] This progression sharply contrasts with the French experience of Justinian's Plague, which did not go further inland than Clermont and Reims, and which quickly subsided.[62]

In Avignon, seat of Pope Clement VI, one of the wisest, most cultivated, and most tolerant of medieval popes, the papal physician Gui de Chauliac, in his *Grande Chirurgie* (*Great Surgery*) gives us perhaps the clearest and most incontrovertible picture of the two distinct forms of the Black Death:

> The great mortality appeared at Avignon in January, 1348, when I was in the service of Pope Clement VI. It was of two kinds. The first lasted two months, with continued fever and spitting of blood, and people died of it in three days. The second was all the rest of the time, also with continuous fever, and with tumors in the external parts, chiefly the armpits and groin; and people died in five days. It was so contagious, especially that accompanied by spitting of blood, that not only by staying together, but even by looking at one another, people caught it, with the result that men died without attendants and were buried without priests. The father did not visit his son, nor the son his father. Charity was dead and hope crushed.
>
> I call it great, because it covered the whole world, or lacked little of doing so. For it began in the East, and thus casting its darts against the world, passed through our region toward the West. It was so great that it left scarcely a fourth part of the people. And I say that it was such that its like has never been heard tell of before; of the pestilences in the past that we read of, so was so great as this. For those covered only one region, this the whole world; those could be treated in some way, this in none.

> For this reason it was useless and shameful for the doctors, the more so as they dared not visit the sick, for fear of being infected. And when they did visit them, they did hardly anything for them, and were paid nothing; for all the sick died, except some few at the last who escaped, the buboes being ripened.[63]

Gui de Chauliac, to avoid dishonor, kept on ministering to the sick despite his admitted terror; near the end of the epidemic he caught the disease himself in its bubonic form, but survived as, after six weeks of continuous fever, the bubo ripened and burst.

The disease, according to de Chauliac's account, came to Avignon from Aix or Marseilles in its pneumonic form—carried from lung to lung. At some point, the disease clearly either passed into the rat population or was transmitted via *Pulex irritans*, because the bubonic form of the disease asserted itself in Avignon after two months of infection. It seems that the good doctor, in any event, suffered from an uncomplicated—even a mild—case of bubonic plague; there is no hint of multiple buboes, skin hemorrhages, or other indications of septicemic plague.

Gui de Chauliac kept his illustrious patient, Pope Clement VI, safer than he kept himself; he instructed the pope to sit alone in his study between two enormous bonfires kept continuously burning; Gui believed that this would cleanse the pestiferous air and keep the pope free of infection. Thus, less by the fires than by the enforced isolation, he kept alive one of the few voices of tolerance and reason of that tortured age. When the populace of France and Germany, looking for scapegoats and poisoners, seized upon the Jews, Clement issued bull after bull declaring their innocence of the charges and insisting that they not be forced into baptism. In Avignon, Gui de Chauliac had made the sensible observation that the Jews, whom, he noted, lived in squalid and filthy quarters, died of the disease as much as anyone else.[64] Under Clement's protection, they were able to build up better quarters, in Avignon and in other papal towns. As one writer puts it, "the Jews repeated with the rest the popular refrain of Clement's day: 'Clement in name! Clement in deed!'"[65]

But the power and influence of this most worldly pope were severely limited, and the aura of his protection did not extend for the Jewish community much beyond the gates of Avignon. Throughout Germany, the Jews were burned by the thousands or banished. In Basle, Freiburg, Gotha, Eisenach, Amstadt, Ilmen, and innumerable other towns they were slaughtered. In Esslingen the Jews themselves gathered in their synagogue and set it on fire. In Speyer some were murdered by a mob, and others, in desperation, set fire to their own homes. In Strasbourg, where the town council tried in vain to protect them, two thousand Jews were burned in their own cemetery.

Avignon had, as well as Gui de Chauliac, another keen observer of the plague. He is an unnamed canon of the Low Countries who was caught in Avignon by the plague while accompanying his cardinal to the curia. His account takes the form of letters he wrote to friends in Bruges.

> The disease is threefold in its infection; that is to say, firstly, men suffer in their lungs and breathing, and whoever have these corrupted, or even slightly attacked, cannot by any means escape nor live beyond two days. Examinations have been made by doctors in many cities of Italy, and also in Avignon, by order of the Pope, in order to discover the origin of this disease. Many dead bodies have been thus opened and dissected, and it is found that all that died thus suddenly have had their lungs infected and have spat blood. The contagious nature of the disease is indeed the most terrible of all the terrors (of the time), for when anyone who is infected by it dies, all who see him in his sickness, or visit him, or do any business with him, or even carry him to the grave, quickly follow him thither, and there is no known means of protection.[66]

The canon goes on to describe two other forms of the pestilence, which turn out to be bubonic plague, with buboes under the arms or in the groin. People are even reluctant to speak with one whose relatives have died, as it has been noticed that in a family where one dies the rest soon follow. More than half the people are already dead, the canon writes: "Within the walls of the city there are now more than 7,000 houses shut up; in these no one is living . . . the suburbs contain hardly any people at all."[67] Eleven thousand people have been buried in a new cemetery the pope consecrated for the purpose; this is in addition to those buried in religious and hospital cemeteries. Meanwhile, four fifths of the inhabitants of Marseilles are dead; the disease stretches across Provence as far as Toulouse in the West. In three months time, from January 25 to April 27, 62,000 bodies have been buried in Avignon, says the canon. It is impossible to know how accurate these numbers are, though one German account substantially confirms the numbers the canon supplies.[68]

After two months the pneumonic epidemic burned out, leaving bubonic disease in its wake. While the two descriptions here don't inevitably suggest transmission from person to person by the human flea, neither Gui nor the canon suggest that the pestilence is no longer contagious. It is also possible that the disease spread both through human fleas and rat fleas; rats may have nibbled on abandoned human bodies and contracted the plague that way, or been exposed to bloody human spittle.

So great was the mortality at Avignon that the pope consecrated two new cemeteries to bury the victims; when those were full, he consecrated the Rhône River so that bodies could be flung into it. It is estimated that half the population died.

The pestilence, in its pneumonic form, continued its long reach westward, toward the Atlantic, and southwest toward Spain. It passed through the gold, green, and stony hills of the Languedoc, to Carcassone, a walled city already a millennium old. Carcassonne, even before the plague, had a tragic history. It had once been the seat of the youthful, gallant Raymond-Roger, Viscomte of Carcassonne and Beziers, and his court, a haven for artists, troubadours, poets. In 1209 the viscomte flouted the power of the French king and the Catholic Church, who wanted to uproot the Cathar heresy from the south of France. The Cathars, tolerant and gentle, opposed the power and the luxury of the Church, and won many converts in the south. The Church, fearing a threat to its power, fought back: the Dominicans and the Inquisition were two Church institutions set up deliberately to deal with the Cathar heretics. The Dominicans tried to win converts where they could; the Inquisition rooted them out more directly; both worked hand in glove to destroy the threat of heresy by any means necessary.

Raymond-Roger, only twenty-four years old and a Catholic himself, publicly offered his sword to the "Jews, Moors, and heretics" of his region. But he could not protect them. The Church declared a crusade against the heretics, the so-called Albigensian Crusade, and thousands of soldiers poured down from the north of France, eager to fight a Crusade so close to home, and to win salvation and booty in the process. The Crusade was, also, the largest land grab in French history. The twenty thousand citizens of Béziers were massacred to the last child, and the Crusaders under the northern French robber baron Simon de Montfort took all that they had owned. They then proceeded to Carcassonne, where they duped the young count into giving himself up to protect his people. He died in the Carcassonne dungeon three months later, and the citizens of Carcassonne were thrust out into the world with only the clothing on their backs.

Later, Carcassonne became a place of torture and terror for the Cathars, as the Inquisition established its seat in the old town, and dragged the heretics out of their remote mountain villages in the Corbières and Pyrenees for "questioning." By 1329, the last Cathar stronghold had fallen, the last heretic consigned to the flames.

As plague entered Carcassonne nineteen years later, the inhabitants were no longer the sophisticated people of the Languedoc, who tolerated different beliefs and customs, and cherished their independence from Church and king. The new inhabitants, descendants of northern land-grabbers and Crusaders, turned upon the Jews, accusing them of poisoning wells, of spreading noxious matter, of deliberately killing Christians with the plague. They dragged them out of their homes and burned them alive in the streets. The slaughter of Jews in Carcassonne and the nearby Narbonne, once home to a thriving, prosperous, and intellectual Jewish community, was "particularly thorough," the German scholar Johannes Nohl tells us in his 1926 book, *The Black Death*.[69]

Meanwhile, the plague spread throughout France, to the north and east of Provence. It reached Lyons near the Swiss border sometime in 1348, as twenty-six wills attributing the cause of death to plague attest.[70] Toward the end of July 1348, it had established itself in Normandy—which means that, in half a year, it had spread the entire length of the country. A manuscript from the abbey of Foucarmont contains a note:

> In the year of grace 1348, about the feast of St. James [July 25] the great mortality entered into Normandy. And it came into Gascony, and Poitou, and Brittany, and then passed into Picardy. And it was so horrible that in the towns it attacked more than two-thirds of the population died. And a father did not dare to go and visit his son, nor a brother his sister, and people could not be found to nurse one another, because, when the person breathed the breath of another he could not escape.[71]

Gilles Li Muisis, the abbot of the Benedictine monastery of St. Giles at Tournai, in an appendage to his earlier entry[72] apparently written in 1350, described the progress of the disease in Tournai in the year 1349:

> I have tried . . . to let future generations believe that in Tournay there was a marvelous mortality. I heard from many about Christmas time who professed to know it as a fact that more than 25,000 persons had died in Tournay, and it was strange that the mortality was especially great among the chief people and the rich. Of those who used wine and kept away from the tainted air and visiting the sick few or none died. But those visiting and frequenting the houses of the sick either became grievously ill or died. Deaths were more numerous about the market places and in poor narrow streets than in broader and more spacious areas. And whenever one or two people died in any house, at once, or at least in a short space of time, the rest of the household were carried off. . . . And certainly great was the number of curates and chaplains hearing confessions and administering the Sacraments, and even of parish clerks visiting the sick with them, who died.[73]

It is hard to imagine a clearer description of the workings of an infectious disease transmitted directly from person to person than this.

In Paris, the clearest record of the Black Death came from the priest Jean de Venette, a professor of theology at the University of Paris, and head of the Carmelite order. He claims that "a healthy person who visited the sick hardly ever escaped death"—and that something on the order of fifty bodies a day[74] were taken from the Hôtel-Dieu. This was not a hotel but a hospital; the sisters there did not shirk taking care of the sick, and "many were called to a new life with Christ." The plague, he says, lasted through 1348 and much of 1349, and left houses, towns, and country places without inhabitants. Li Muisis of Tournai, also, spoke of deserted houses, cattle wandering through fields unattended, barns and wine cellars left open to the wind.

The Black Death came to Germany, to Spain, and the stories are substantially the same, except that in Germany the horror induced by the pestilence gave rise to a strange and terrible new brotherhood—the Flagellants, the Brethren of the Cross. Many devout Christians looked on the plague as God's just punishment for their sins; some reformed their ways, married their concubines, abstained from drinking and gambling, and appeased the friends and neighbors they had insulted or offended.[75] But the Flagellants took self-abasement to an extreme. The roots of the movement seem to lie in Eastern Europe, but it was in Germany that it spread and flourished. Two or three hundred men, with women sometimes in the rear, would, two abreast, form a snakelike chain behind their leaders, who carried banners of purple and gold. They wore dark clothes adorned with red crosses; they hid their faces in cowls, and trudged along with their eyes fixed firmly on the ground, in a silence that they would sometimes break with dirges. They walked from town to town, and as they approached each new city the bells would be rung at word of their approach, and the inhabitants rush out to greet them. Sometimes they would hold their liturgies in church; more generally in a square or market, surrounded by a gaping multitude. There they would cast off their coarse outer garments down to the waist, leaving only a thin linen cloth for cover from waist to feet. After parading around in a circle, they would stop abruptly, at a signal from their leader, and fling themselves to the ground in various penitential postures; their leader would stride about the huddled penitents, beating them for their sins. Eventually the entire group sprang up again, each man seizing his own scourge—a stick with three tails with large knots from which sprouted tiny iron spikes. They whipped their bare backs until the blood flowed, and their frenzied shrieks and prayers went up, they devoutly hoped, to heaven. Some died, and this would lead to an extra performance in honor of the deceased. These edifying demonstrations went on all across Germany and into France, and were soon alloyed with vicious expressions of hatred against the Jews, until Pope Clement VI put a stop to it all, threatening the marchers with excommunication.

How much these determined marchers, wending their way through Germany and down to France as far as the pope's seat in Avignon, succeeded in spreading plague we cannot know.

This wise and thoughtful pope made one tragic misjudgment. Remembering, perhaps, the great procession with which Pope Gregory the Great was said to have stopped the plague in Rome, Pope Clement called his faithful to Rome in 1350 to celebrate a jubilee. The thousands of pilgrims, crushed together in the narrow streets of Rome, set off another conflagration: plague broke out among them with renewed ferocity, and ravaged all of Italy once again.[76]

* * *

The English king Edward III, who in 1337 had fired the opening salvos against the French in a war that was to last over a hundred years, may have deceived himself, in 1347, into thinking that God had no high opinion of the French. A year earlier, he had won a stunning victory over Philip VI, the feckless French king, in the famous Battle of Crécy, where the English archers with their longbows decisively routed the French crossbowmen, and the French army, shocked and pounded into a disorganized rabble by Edward's horsemen, was mulched to ribbons.[77] Next the Black Death, deadlier than any English army, began to overrun the country.

But Edward had little time to rejoice in his enemy's suffering. In August 1348, his own daughter, the fifteen-year-old Princess Joan, on the way to her wedding with King Pedro of Castile, died of the plague in Bordeaux. By then, as well, the Black Death had already landed in Dorset.

Or did it first come to Bristol? No one knows for certain, but it seems that the plague did not begin its death march across England until the summer of 1348, when France was already reeling. Most likely the disease was brought to England on a ship from Calais or from Gascony, both then under English rule. It was much easier for people and goods to arrive in England than to travel overland across the Alps or across continental Europe. England lay at the center of a vast trade web that spanned Europe from the Baltic to the Mediterranean. Though the Black Death came later to England, come it must, and come it did.[78]

The story in England is the same as that for the continent. Of the contagious spread of the disease, which has so bedeviled those who have tried to force the rat flea/bubonic plague model onto the Black Death in England, there can be no doubt. A sharp demonstration of this contagious spread appears in Graham Twigg's *The Black Death: A Biological Reappraisal*, the first of the plague skeptics' manifestos.[79] In a table that compares the rate of spread during the Black Death to that of rat-borne modern epidemics, Twigg shows that the difference is, indeed, quite striking. From Melcombe to Bristol in 1348, the rate is either 1.5 or 4 miles a day (Twigg does not explain the discrepancy), while the plague from Bristol to London moved along at the brisk clip of two miles per day. In contrast, in South Africa's outbreak of 1899–1925, rat-borne plague slowly expanded at the rate of eight to twelve miles *per year*; while in the 1906 outbreak in Suffolk, England, the plague is even more sluggish, moving at a rate of only four miles per year.[80]

As Twigg recognized, there is indeed something very unusual about the Black Death compared to modern, and much studied, plague outbreaks. The British Isles have been a particularly fertile field for the plague skeptics, as they have produced incomparable death records. With England, we are on much firmer footing than on the Continent about how many died and where. Here, we can really see the rate of spread; and so the plague skeptics have mostly used the English records to argue that the germ which caused the Black Death could not possibly have been *Yersinia pestis*.

Unfortunately for Twigg's credibility, he proposed anthrax as the agent instead—and anthrax, which is only infectious in spore form, has no capacity for person-to-person transmission. In 2001, two epidemiologists from the University of Liverpool, Susan Scott and

Christopher Duncan, published a book called *Biology of Plagues: Evidence from Historical Populations*, which makes a similar argument using mathematical modeling.[81] They go so far as to assert that "[*Yersinia pestis*] was certainly not the causative agent in the Black Death," and, even more baldly, when speaking of the famous plague outbreak in the English village of Eyam, Derbyshire, in 1665, "*It is a biological impossibility that Yersinia pestis was the causative agent*" (italics mine).[82]

Many priests died of plague during the Black Death, though how many died, and how many deserted, we have no way to know. One scholar estimates that, overall, approximately 40 percent of clerics died of plague in England.[83] This does not imply that clerics provide us with a representative sample of the overall death rates in the population, which may well have been lower: along with doctors, clerics—who administered last rights to the dying—suffered the most from exposure to contagious plague. That infection might have come either from a bite from a human flea infesting the linens, or from the expectorations, even the speech, of the infected patient as the priest or doctor bent over him.

Though the ecclesiastic records, and the manorial court rolls, give us imperfect data, they do seem to supply us with an indication of where plague was raging, and when. The rate of spread, a rate reported at between one and ten miles a day, probably was as fast as a man could journey over the rough roads of medieval England. Stricken with pneumonic plague, a man can travel for a day or two before the disease overwhelms him. People fleeing plague often took it with them—either among their layers of clothing or their goods, or incubating in their lungs. The transit too rapid for infected rats was not too rapid for human fleas and for pneumonic infection carried by travelers themselves.

Plague came to Ireland, too, by sea, and probably from Bristol.[84] There are poorer statistical records from Ireland than from England, but there is an impassioned record left by Friar John Clyn of

Kilkenny. Despite the late date (he apparently wrote toward the end of 1349), we see from his account that the Black Death had lost none of its particular virulence or transmissibility:

> From the beginning of all time it has not been heard that so many have died, in an equal time, from pestilence, famine, or any sickness in the world. . . . The plague too almost carried off every inhabitant from towns, cities and castles, so that there was hardly a soul left to dwell there. This pestilence was so contagious that those touching the dead, or those sick of it, were at once infected and died, and both the penitent and the confessor were together borne to the grave. Through fear and horror men hardly dared to perform works of piety and mercy; that is, visiting the sick and burying the dead. For many died from abscesses and from impostumes and pustules, which appeared on the thighs and under the arm-pits; others died from affection of the head, and, as if in frenzy; others through vomiting of blood. . . .
>
> The pestilence raged in Kilkenny during Lent, for by the 6th of March eight friars Preachers had died since Christmas. Hardly ever did only one die in any house, but commonly husband and wife together, with their children, passed along the same way, namely the way of death.
>
> And I, Brother John Clyn, of the order of Minorites, and the convent of Kilkenny, have written these noteworthy things, which have happened in my time and which I have learned as worthy of belief. And lest notable acts should perish with time, and pass out of the memory of future generations, seeing these many ills, and that the world is placed in the midst of evils, I, as if amongst the dead, waiting till death do come, have put into writing truthfully what I have heard and verified. And that the writing may not perish with the scribe, and the work fail with the labourer, I add parchment to continue it, if by chance anyone may be left in the future

> and any child of Adam may escape this pestilence and continue the work thus commenced.[85]

Brother John Clyn wrote only two words more "*magna karistia*"—"great dearth." Then, on the parchment he left for posterity, there is another note in a different hand, "Here it seems the author died."[86]

The Black Death in the British Isles killed the great as well as the small. While only Princess Joan in King Edward's family died, three archbishops of Canterbury perished in quick succession, at least two of them of plague. The poet Richard Rolle died of it, as did the great philosopher William of Ockham. Entering monasteries, the plague slew abbott and lay brother alike. The students of Oxford died in their thousands, and were buried right at the university, in a plague pit dug into a part of New College—the garden or the cloister—left deserted by the plague.[87]

* * *

The Black Death was a very different disease than the Justinian Plague, even though the same germ caused both of them.[88] Unlike the First Pandemic, which only affected places within reach of Justinian's restored web of trade and commerce, the Black Death moved inland with astounding rapidity, a rapidity too great for the movement of rats or even for rat-flea-infected goods.[89] Furthermore, that the disease was clearly contagious is demonstrated by the extensive records left by careful observers. This is in complete contrast to the sources for the Justinian Plague, none of whom mention contagion—except for Paul the Deacon, who lived two hundred years after the Ligurian plague outbreak he describes. Procopius, in fact, explicitly denies that those who cared for people with the plague contracted it; nothing, given both the written testimony and the records of deaths among priests and physicians, could be further from the truth for the Black Death.

Pope Clement attempted to collate the reports he was given for mortality across the known world, and arrived at a figure of 23,840,000 dead out of 75 million inhabitants. This is a full 31 percent.[90] A nineteenth-century scholar, adding up whatever figures he could find for all of Europe, estimates that at least one quarter of the entire population died of the Black Death, some 25 million people.[91]

That there was something very different about the Black Death from the usual *pestis inguinaria*, or Oriental Plague, was recognized by two early writers on the Great Mortality.[92]

J. F. C. Hecker, a nineteenth-century German scholar who had thoroughly reviewed original texts in many languages, wrote the first general description of the Great Mortality. He noted:

> The descriptions which have been communicated contain, with a few unimportant exceptions, all the symptoms of the oriental plague which have been observed in more modern times. No doubt can obtain on this point. The facts are placed clearly before our eyes. We must, however, bear in mind that this violent disease does not always appear in the same form, and that while the essence of the poison which it produces . . . remains unchanged, it is proteiform in its varieties, from the almost imperceptible vesicle, unaccompanied by fever, which exists for some time before it extends its poison inwardly, and then excites fever and buboes, to the fatal forms in which carbuncular inflammations fall upon the most important viscera.
>
> Such was the form which the plague assumed in the 14th century, for the accompanying chest affection which appeared in all the countries whereof we have received any account, cannot . . . be considered as any other than the inflammation of the lungs . . . a disease which at present only appears sporadically, and . . . is probably combined with hemorrhages from the vessels of the lungs.[93]

Despite the nineteenth century language and Hecker's failure to understand the infectious causation of disease, this description captures the mystery in its essence. Here is the familiar Oriental Plague, but subtly altered, transformed so that in every country it appeared as a contagious lung disease, a form only sporadically seen in Hecker's time, or our own.

Cardinal Francis Aidan Gasquet, an English scholar of the late nineteenth century, makes the same distinction between the "ordinary Eastern or bubonic plague" and the Great Mortality. In its common form, the disease shows bubonic swellings and carbuncles under the arms and in the groin:

> In this ordinary form it still exists in Eastern countries. . . . The special symptoms characteristic of the plague of 1348–9 were four in number:
>
> 1. Gangrenous inflammation of the throat and lungs;
> 2. Violent pains in the region of the chest;
> 3. The vomiting and spitting of blood; and
> 4. The pestilential odour coming from the bodies and breath of the sick.
>
> In almost every detailed account by contemporary writers these characteristics are noted. And, although not all who were stricken with the disease manifested it in this special form, it is clear that, not only were many, and indeed vast numbers, carried off by rapid corruption of the lungs and blood-spitting, without any signs of swellings or carbuncles, but also that the disease was at the time regarded as most deadly and fatal in this special form.[94]

Plague skeptics, beginning in 1984 with Twigg, protest that pneumonic plague does not have the potential for spread exhibited by the Black Death; it moves too slowly and is too little infectious to be a candidate for the agent of this greatest pandemic in human history. What the plague skeptics do not seem to understand is that the type of plague makes all the difference. The plague germ that circulates among the marmots of Central Asia cannot be compared, in virulence and transmissibility, to the plague that kills prairie dogs in their millions, or which sometimes enters human populations through fleas and rats. As Suleimenov, the Kazakh plague specialist, insists, marmots can catch and spread pneumonic plague amongst themselves—they have been found dead in the wild with bloody froth on their mouths and noses. Furthermore, he says, marmot plague strains have a pneumotropism—they move immediately to the lungs, in marmots or in humans.[95]

What is inarguable, though, is the repeated twentieth-century experience with violent outbreaks of pneumonic plague in Manchuria, epidemics that killed thousands, and that have been repeatedly traced back to the hunting, skinning, and eating of sick marmots.

The scientists who know best the deadly power of plague—those Soviet scientists who turned plague into the world's most lethal bacterial weapon—made that weapon from these same ancient marmot strains, which, set loose once before, seem to have wiped out perhaps a quarter of humanity.

Marmot plague, however lethal to human beings, is still marmot plague; it has a long evolutionary history in the marmot population. The Black Death, on the other hand, must have evolved away from the original strain in a direction that favored transmission among humans. The Black Death became, in a limited, short-term sense, a human disease: much of it spread lung to lung, some of it through the unpleasant intimacies of *Pulex irritans*, with which our not always cleanly ancestors were surely only too well acquainted. Perhaps, sometimes, rats and rat fleas passed the disease on as well, and no doubt they played the principal role in converting plague into a constant, if somewhat less virulent, menace over the next several centuries. But certain peculiarities of the Black Death suggest that the disease agent had undergone some sort of evolution. The ubiquitous "pestilential odor" of the medieval records, for one thing, is never mentioned in the modern clinical literature on plague. As Wu Lien-teh says, "The majority of modern authors . . . agree that they could not perceive any characteristic smell in plague sufferers, although much was made of that sign by old recorders."[96] Wu mentions, also, that the apparent "necrosis of the lungs" described by Simon de Covino has also been noted by modern observers, though rarely.[97]

These unusual characteristics could have been caused by the long chain of human-to-human transmission, as the Black Death evolved from marmot plague to a human disease. As Wu puts it, "The fact that the Black Death does not quite correspond to the forms of infection as it is known today cannot eliminate the ample evidence that it was plague. The descriptions of both the bubonic and the pneumonic type, as given by contemporary observers, leave no room for doubt."[98]

Despite this apparent evolutionary process, the Black Death never became a permanent human specialist like smallpox. It may have been simply too virulent. Though some people survived the bubonic form of the infection, all the commentators insist that blood-spitting was invariably, and rapidly, fatal. In order to become a permanent human affliction, pneumonic plague probably would have had to lose something of its virulence. Even smallpox, the most virulent of the highly contagious human diseases, kills no more than 30 to 50 percent of the population.[99] On the other hand, bubonic plague—the "mild," ordinary *pestis inguinaria*, kills 50 to 60 percent of those infected. It sustains itself because of its reservoir host, not because it passes from person to person.

Still, the pneumonic phase of the Black Death lasted for a number of years, from no later than 1339 until 1351 and perhaps beyond. This raises an interesting problem. How did so lethal a disease manage to keep spreading from person to person? This persistence may have something to do with the stealthy nature of the infection. Some diseases, such as measles, are especially infective because people transmit measles germs through sneezing and coughing while they are still walking around—before the disease has had a chance to make them sick. The plague germ seems to have a reverse strategy. It produces no symptoms at the beginning, and victims can travel while it does its work. The terror the disease produced was the handmaid of the infection: people fled the pestilence on ships, on wagons, on horseback, and spread the disease wherever they went.

* * *

Eventually humanized plague died out, though plague was not yet finished with the peoples of Europe and Asia. It would repeat its assaults in wave after wave, reducing the population, changing labor patterns and economic relations, as well as, perhaps, more subtle things—art, religious views, the political order. But later plague was not the same as the Black Death. It took many thousands of lives, and generations grew up in its shadow. But the Renaissance plague was a tribulation, not a holocaust.

The Black Death, the Great Dying, was the most terrible pandemic in all of human history. Its murderous tour of Europe and Asia has never been repeated, by plague or by any other disease. Natural plague has never since had the opportunity to go pneumonic and spread out from Central Asia across the earth.

If the Black Death comes again, it will need human assistance to do so.

# V

# THE RENAISSANCE PLAGUE

**Second Carrier**
**I think this be the most villainous house in all London road for fleas: I am stung like a tench.**

**First Carrier**
**Like a tench! by the mass, there is ne'er a king christen could be better bit than I have been since the first cock.**

*HENRY IV*, PART I

View of the manner of burying the dead bodies at Holy-well Mount during the dreadful plague in 1665.

COLOR ENGRAVING BY S. WALE

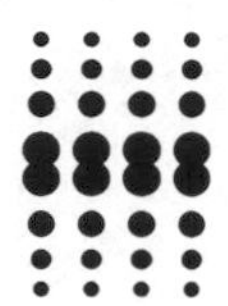

The Black Death, a hot wind out of Central Asia, scourged the known world for several years and then vanished. But plague itself would take centuries to vanish from Europe, and still longer from the Middle East. What changed, in the series of plague epidemics that washed over Europe and Asia over the next several centuries, was the peculiar constellation of symptoms that had marked out the Black Death from all other plague epidemics. This strange admixture, from the "plague tokens"—hemorrhages under the skin considered the infallible precursors of death—to bubonic swellings, to blood-spitting and rapid death, are not found in either earlier or later epidemics on a large scale. In later accounts of the disease from Renaissance England, for example, or from Italy, there is no mention of blood-spitting, though the tokens and the buboes remain. In the great Manchurian epidemics of the early twentieth century, there are few mentions of buboes—the epidemics are almost exclusively pneumonic in character.

The pneumonic phase of the Black Death eventually died out, and plague in Europe and the Middle East became almost exclusively vector-borne. But so lethal a disease, however it is transmitted, cannot remain for ever in the same vicinity. It moves on or dies out—having killed or immunized everyone exposed to it. Though recovery from pneumonic plague is rare, about 40 percent of bubonic plague sufferers in the pre-antibiotic era managed to survive it; in a population that suffered from the mixed infection of the Black Death, therefore, it is probable that there would have been more survivors who had acquired immunity. Furthermore, it is now evident that silent plague infections are actually relatively common. Though this was long denied by plague experts, recent research in Madagascar and Ecuador shows that the plague germ causes a rather high rate of inapparent or asymptomatic infections (8.4 percent of the exposed population, according to one report,[1] 13.5 percent in another[2]). For reasons that are poorly understood, some people who are infected with the plague germ do not fall ill, though analysis of their blood yields evidence of seroconversion—the presence of antibodies to particular antigens present in plague.[3] This silent infection acts like a vaccine, and, for a time at least, the seroconverted person is immune to plague. Combined with the people who recovered from the bubonic form of plague, these seroconverted people helped reduce the field of possible targets. Returning waves of plague disproportionately affected children, who had not yet been exposed.[4]

At some point after 1351, then, plague in Europe changed its character. References to blood-spitting vanish from the records. Now the buboes, carbuncles, and skin hemorrhages (ecchymoses) known as the plague tokens are the plague symptoms we read about. These are symptoms of vector-borne disease—either rat- and rat-flea-borne, or human-flea-borne transmission—a very different pattern from that of the Black Death. Plague transmission becomes largely seasonal—the plague epidemics in Renaissance England, for example, broke out in the summer, whereas the Black Death in one form or another raged all year long. The Oriental rat flea, *Xenopsylla cheopis*, cannot be active in cold weather.[5]

* * *

The slow pace of the next waves of epidemic plague suggests the presence of rat-borne disease; the quick spurts of transmission among people in the same household, or those sharing beds and clothing, point to the role of *Pulex irritans*. There is every reason to think that both forms of transmission played an active role in the plagues of the European Renaissance. Some people, including Twigg, have argued that there were no rats in late medieval England. This is patently false. That there were rats in Europe cannot be questioned: the art of Hieronymus Bosch, the portraits of ratcatchers, the legend of the Pied Piper of Hamelin, which dates from around 1280, all indicate that the house rat was a common denizen of Europe by the late Middle Ages.[6]

The Black Death might have burned out, but Europe had only a few short years to recover before waves of plague began rolling over the Continent and the British Isles again. Some scholars have maintained that the true demographic destruction, the reason for the failure of Europe's population to recover for centuries, instead of decades, came from these repeated waves of infection. In England, the next epidemic came in 1360, which killed, according to historian Josiah C. Russell, 22.7 percent of the population. This was followed by the epidemic of 1369, killing 13.7 percent and again in 1375, with a mortality of 12.7 percent.[7] Since many people who survived the Black Death had immunity either from the disease or from subclinical infection, the second epidemic affected mostly children, a familiar pattern in the history of infectious disease. (Smallpox, measles, diphtheria, and other infections became, over time, mostly diseases of children.[8]) This had an even more depressing effect on the population. A similar pattern of repeated infection also characterized the Continent.

Little is known about these early post–Black Death plague outbreaks. But as we move closer in time to our own era, the record becomes clearer. By 1500, there is little doubt that epidemics of plague were washing over Europe. The timing, the seasonality, the slow creep from neighborhood to neighborhood, city to city, the plague's clear preference for the haunts and dwellings of the poor, all distinguish these later outbreaks from the democracy of death that characterized the Second Pandemic. It is not for nothing that the final terrible burst of plague in England was known as the

Poore's Plague. The same thing might be said, in general, to characterize all the repeated waves of plague from the late fourteenth century until, at last, the plague vanished from Europe.

In the year 1610 a poor man named Dobson had a lodger who died of plague. Lodgers typically lived in cellars, dank, dark, and overrun with rats. They were the poorest of the poor in England, and the frequent targets of cleanup campaigns to move them out of their parishes and to lower the risk of plague. Dobson took the lodger's coat, which he did not bother to wash or even air, and sold it to a townsman named Wexe, who lived in Westerham. Wexe died of plague. Then several people who prepared Wexe's corpse also died of plague.[9]

The outbreak ended there. Westerham had recently suffered a major bout of plague; perhaps the disease lingered in a few infected rats who may have crossed paths with the unfortunate lodger. But the quick succession of plague cases arising from contact with the lodger's coat or with Wexe's body suggests that *Pulex irritans*, nestled in the unwashed coat, had caused the outbreak.

Fleas infested the clothing and beds of more than just the poor, and it was not uncommon for the well-to-do as well as the poor to stay in inns with shared beds and unwashed linen, as often as not infested with fleas. Samuel Pepys, the famous diarist, shared a bed in Portsmouth with a friend and wrote that "all the fleas came to him and not to me."[10] But fleas are not often mentioned in the literature of the time,[11] any more than we write much about carpenter ants or spiders today; they are part of the mundane backdrop of our lives, and we take them for granted. It is difficult for us to imagine that our not so remote ancestors were plagued with parasites of all kinds that they also took for granted: pinworms, head lice, *Pulex irritans*. *Pulex* is as old as time: the recent discovery of the so-called Iceman on an Alpine ridge near Innsbruck, Austria, revealed that even this five-thousand-year-old man was plagued with *Pulex irritans*—two were found on his body.[12] *Pulex* also survives and thrives in a much wider temperature range than does the Oriental rat flea.

The plagues in England and the rest of Europe during the sixteenth and seventeenth centuries, unlike the Black Death, were predominately seasonal. The deadliest season for plague was the summer, though in several instances, at least, the disease only died down in winter to a lower level, without disappearing. Another striking feature of these later plagues is their affinity for crowded urban areas, and their slow movement from one neighborhood to another. The epidemics were characterized by rapid bursts of explosive plague within a confined area, followed by a slow drift, hesitant, irregular, and terrifying, across a city or a region.

About contagion, too, there can be no doubt—these later plagues spread person to person, though not in the way of the Black Death. There is no mention of blood-spitting, and no one speaks of contagion through a single glance. Contagiousness appears within families and households, among people sharing clothing or beds, and, saddest of all, among those who, because of the perceived necessity to stop the spread of the disease, are nailed up together in their own homes to die. Despite the clear presence of the human flea, and the short chains of contagion that linked people through common clothing or common domiciles, there can be no doubt that rats and rat fleas played a key role in the spread of these English epidemics. Not only their seasonality, but also the way plague cases clustered in the filthy alleys of the poor tells us that rats must have brought plague into those streets.

England, and the rest of Europe (except for the Caucasus and, possibly, certain regions around the Volga), have never been home to the kind of burrowing rodents that could become natural foci for plague. Even the marmots once widespread throughout Europe are solitary creatures, which, since they do not live in colonies, cannot become permanent reservoirs of plague. Yet, plague seems to have become endemic in black house rat colonies for a time, much as it does among rodents in the wild. There were only a few short periods between 1485 and 1665 (1612–1624; 1654–1664) when there were no apparent human plague cases somewhere in the country.[13]

The large-scale epidemics that swept over England so many times in the sixteenth and seventeenth centuries do not seem to have originated in this smoldering native plague. One writer calls the epidemic waves instead "the end result of broad tides of infection moving from the eastern Mediterranean across the whole of Europe."[14] As in Justinian's Plague, and the first days of the Black Death, port cities were the most vulnerable. The familiar picture emerges: the docked ships from overseas, the flea-infested goods, the infected rats darting off ships to disappear among the native rodents of the port. Furthermore, the most vulnerable places were those ports closest to the continent: Bristol, Norwich, Exeter. London, too, with its river ports, its dense population, and the filthy, rubble-strewn streets that lay beyond the City walls, was the natural target, and inevitable destination, for some of the greatest epidemics of all. In short, foreign rats brought far more virulent plague, while domestic rats only infected handfuls of Englishmen.

During this period, in general, plague passed by the small villages and the countryside and hunted in the denser fields of urban areas and crowded suburbs. The City of London was far wealthier and more orderly than the areas that surrounded it, and suffered much less from plague. Like Paris, London had a belt about it of the indigent, the criminal, the cast-off, and the forgotten, where the disease found more hospitable ground. Plague favored, also, the newly industrializing towns. England was coming to dominate Europe in its production of woolen textiles: the sites of textile production and drapery manufacturers quickly became towns with textile workers, many of whom were desperately poor. And these densely populated, impoverished towns were fertile fields for plague. Even so, no areas were immune, and the unpredictable ways that plague struck this town, then that village, and left a third untouched, must have added to the fear of death the terror of the inexplicable.[15]

From the distance of several centuries, though, we can see through the randomness to perceive clear if somewhat irregular patterns. Plague went where rats were most at home. It diffused slowly from one center of infection to another, except for the rapid jumps from household to household when someone brought in infected woolens from a mill, or clothes from a dead man. While there were regulations in force, as we shall see, as to the disposition and destruction of clothes and bedding from plague victims, these were often disregarded, as people shivering in the damp chill of a London winter might risk plague to have a heavy blanket or warm coat. Rat-borne plague drifts from place to place, slowly moving from one rat-infested area to another—as the rats die, their fleas move to other rats, and so on. But *Pulex*-borne plague jumps from place to place, making new little nodes of infection in whatever new household it enters. Blanc and Baltazard put it this way: "Any human carrier of parasites who dies of plague creates around himself a node [*foyer* in the original French] which will be all the more significant the more numerous the parasites are, since these [they mean *Pulex*] will automatically return to people."[16]

Rat plague is like a slowly rolling wave curving along a beach; *Pulex*-borne plague more like a small sharp blaze that may burn for a while and spread, or which may be quickly extinguished. Both modes of transmission are necessary to explain the movements and depredations of plague in Renaissance Europe. Even Blanc and Baltazard, who have done more than any other researchers to insist on *Pulex* as an agent of epidemic plague, admit that in the Europe of the fourteenth to seventeenth centuries the "plague came with the rat, and sporadically touched man through the rat flea."[17] But then, they maintain, the real work among people was done by the human flea.

"The plague in man found then the most perfect conditions for its expansion, thanks to the intense parasitism on the populations of this epoque."[18]

The Great Plague of London, in 1665, was the last and severest of a series of Great Plagues: 1603, 1625, 1636, and 1665. There were other, smaller outbreaks; in fact, in the seventeenth century hardly a year went by without a scattering of deaths attributed to plague. This does not mean that all those deaths were from true plague: "the searchers of the dead" who identified the bodies were not trained physicians (even for those times) but impoverished, ignorant old women who trudged from house to house and examined corpses to certify the cause of death. Still, one expert states that out of the sixty-four years between 1603 and 1666, there were twenty-eight years where plague, beyond dispute, was present in the city.[19]

Unfortunately, despite several centuries' experience, Londoners had learned nothing about how to keep the plague out of their city. There were regulations, mostly pertaining to quarantines and the seizure of suspect goods. But perhaps the scattered mild outbreaks, the occasional pauper dead in the streets, the tiny flares of disease that took only a handful, and those the poorest of the poor, had lulled Londoners into complacency. They had forgotten what the plague could do.

The roots of the Great Plague lay not in the smoldering plague scattered among the city's rats; it lay in the Ottoman Empire, struck by plague in 1661. From Turkey, the plague moved to Holland, with which England was at war; many Dutch sailors were taken as prisoners of war and housed on English soil.[20] Plague's advent in England, in hindsight, was inevitable. But when the full fury of the plague hit, as Walter George Bell points out in his elegant history *The Great Plague of London*, the Londoners, who should have been watching more closely as the plague killed 35,000 in Amsterdam,[21] were utterly unprepared, "the bulk of the people stolidly indifferent."[22] When the plague fell once again upon London it fell on fertile ground: a filthy city, flea-laden, rat-haunted, ripe with ordure.

There were no sanitary regulations, or none that could cope with filth and poverty of London's magnitude. The population of London was about half a million;[23] some 130,000 lived within the City walls and surrounding neighborhoods, and many more outside, in the crowded outer parishes. Streets and alleys were thick with mud; chambermaids emptied pots of urine and nightsoil directly into the streets or into the streams, like the Fleet which ran through the city, "no better," Bell tells us, "than open sewers."[24] Garbage moldered in huge heaps; the streets periodically were swept of refuse, composed of everything imaginable, which was placed into stalls right in the city and left to rot. There were five pesthouses, with space for six hundred plague patients: this was hopelessly inadequate.

Even worse, the first response of London authorities was to deny what was upon them, both for economic reasons and out of the natural desire to pretend that, despite the mounting death toll among the poor, life continued on its normal course. "In part deceived themselves, they cannot be acquitted of a share in wilful concealment, which fear of disorders alone can have excused," writes Bell of the city leaders.[25] Still, the scattering of plague cases in London was sufficient to suspend Parliament in March, its worthies fleeing London for the country. Following hard on their heels went the court, and then the merchant classes. Isaac Newton, who had recently finished his Cambridge studies, left London for his mother's farm in Lincolnshire, where he commenced his *annus mirabilis*, during which he would invent calculus in the quiet of the countryside. In fleeing early, Newton and the others saved themselves; had all of London stayed in place, the death rate would have been much higher. But the poor were left alone, to the fleas and rats and pestilence spilling like a slow tide across the city.

The outbreak began in a poor out-parish, St. Giles-in-the-Field, and spread outward, Bell tells us, "like the points of a star, arrested where meadows lay north and west, but following unerringly the lines of population."[26] Until June the disease spread wholly outside the city walls, in the western parishes, but as the summer progressed and the heat intensified, the pace of the spread picked up, until in July the disease was far more widespread. By the time the outbreak had run its course, there were perhaps only four out of 130 parishes in London that had been spared the plague.[27]

The poor, and other people who did not flee the city when they could, were often highly motivated to deny the presence of plague in their midst. The heartless regulations in force for generations decreed that any household harboring a plague case was immediately to be shut up, with the doors and first-floor windows nailed closed with wooden planks. A red cross was painted on the door, the sign of plague, along with the plangent supplication "The Lord have mercy upon us." A watchman was posted outside. These households became charnel houses: shut up together, not infrequently all the inhabitants died as the plague fleas spread. To be boarded up like this was tantamount to a death sentence, and it is not for nothing that people hid their sick and their dead. The twentieth-century British scholar J. F. D. Shrewsbury quotes J. Noorthouk, an eighteenth-century historian, on the onset of the plague:

> At the beginning of the disorder there were great knavery and collusion in the reports of deaths; for while it was possible to conceal the infection, they were attributed to fevers of all kinds, which began to swell the bills [Bills of Mortality]; this was done to prevent their houses being shut up, and families shunned by their neighbors.[28]

The parish provided a watchman to stand outside each plague house, to prevent anyone from going out or in, and to see to the family's wants. Unfortunately these watchmen came from the lowest levels of the social order; they were frequently drunk, boorish, and indifferent to the welfare of the families they were supposed to protect. The watchman was supposed to purchase and convey food and medicine to the shut-ins, to be their living link to the outside world. In reality, the plague shut-ins suffered from terror, isolation, hunger, and want. In the words of a contemporary anonymous writer:

> No drop of water, perhaps, but what comes at the leisure of a drunken or careless halberd bearer at the door; no seasonable provision is theirs as a certainty for their support. . . . They are compelled, though well, to lie by, to watch upon the death-bed of their dear relation, to see the corpse dragged away before their eyes.[29]

If the family starved to death, the watchman still earned his meager wages all the same. And as frightful as this indifferent guard of their welfare posted outside their door must have been, how much worse the ghoulish old women of the night, the "searchers of the dead," who would enter their houses to certify the dead, determine the cause of death (so far as they could) and to see what bodies needed to be taken away? Then there were the night carts, with their loud rattling on the cobblestones, and their terrifying cry of "Bring out your dead!" If no one remained alive, the entire family would be thrust in a heap onto a death cart, arms and legs dangling, mouths agape, unshrouded, unregarded, lying exposed as the cart lurched onward. If there were any survivors, they had to watch their loved ones—old people, wives, husbands, children, even infants—thrown on the wagon and hauled off to the plague pits, to be thrown in like the forgotten dead of Constantinople, of Florence, of Avignon, of all who had died in the plague pandemics before them. So many were thrown in, in such monstrous heaps, that some of the plague pits in London are still visibly raised to this day.

Worse than the careless watchmen, worse than the searchers, were the plague "nurses" sent in to care for the plague-stricken, who far better had been left alone. These nurses are not to be confused, Bell reminds us, with the "sisters" who for centuries had cared for the sick in hospitals. Like the searchers, these nurses were taken from the lowest stratum of the social order, for no one else would take on such a dangerous task, or work for such a pittance as they were paid by the parish. They supplemented their meager earnings with whatever they could steal from the dying. The plague-stricken had no power to turn the women away, and had to endure their less-than-tender ministrations. Bell relates the observation of one contemporary observer, the Reverend Thomas Vincent, who declared that "the stricken Plague victims were more afraid of the nurse than of the Plague itself."[30] Bell also quotes Dr. Hodge, a skilled physician who unlike many of his colleagues did not desert London in this black hour, on the nature of these women and their effects on those nailed into quarantine:

> But what greatly contributed to the loss of people thus shut up was the wicked practices of the nurses, for they are not to be mentioned but in the most bitter terms. These wretches, out of greediness to plunder the dead, would strangle their patients, and charge it to the distemper in their throats. Others would secretly convey the pestilential taint from sores of the infected to those who were well. Nothing, indeed, deterred these abandoned miscreants from prosecuting their avaricious purposes by all the methods their wickedness could invent. Although they were without witnesses to accuse them, yet it is not doubted but Divine Vengeance will overtake such wicked barbarities with due punishment.[31]

The theory of quarantine was that the only way to prevent the disease from spreading was to shut away all of those infected with it, together with those who shared a household with them. This was the best possible way to ensure that those human fleas infesting the sick would find someone else to jump to. Pesthouses would have served far better as a means of separating the infected from the well, but such pesthouses as London had were full. By the end of May 1665, seven hundred had died, and the number of sick was rapidly rising.

Samuel Pepys, the famed journal-keeper and Secretary of the Admiralty, saw his first plague houses on June 7, and hurried on by, in distress:

> This day, much against my will, I did in Drury Lane see two or three houses marked with a red cross upon the doors and "Lord have mercy upon us!" writ there; which was a sad sight to me, being the first of the kind that, to my remembrance, I ever saw. It put me into an ill conception of myself and my smell, so that I was forced to buy some roll-tobacco to smell to and chaw, which took away the apprehension.[32]

Tobacco was thought to possess protective powers against the plague; schoolboys at Eton were forced to smoke, and those who disobeyed were flogged.[33]

Pepys lived in the walled City, which was much less affected by the plague than the outlying parishes. But life began rapidly to change for everyone: on June 5, all public places of entertainment, such as theaters, were closed down; by the middle of the month, the king and the court had fled. By the end of June, the Bills of Mortality, which listed all the dead in the city—at least in theory—listed nine thousand as dead of plague. These figures were certainly underrepresented: not only did many households, out of fear of quarantine, hide the cause of death, but Jews and Quakers were not included in the rolls, as they buried their own dead in their own cemeteries.

The death rate rose steeply after that, as the summer heightened and the plague continued its irregular march across London. For people like Pepys, the plague was an inconvenience and a source of dread and dismay—his own doctor died of it, and so did Pepys's Aunt Bell. But in the main, his life—and many other lives in the walled City—went on, in the usual tracks of business and pleasures.[34] Their lives were altered by the lack of public entertainments and the absence of the court and many of their peers, and perturbed by frequent sightings of plague houses and death carts, but it was life all the same, and Pepys was giddy with the pleasure of arranging a wedding for important social contacts.

Still, he was "fearful of death . . . in this plague time."[35] By September, Pepys learns that though the Bills of Mortality, giving the death toll for the week, have dropped a little, the plague deaths have increased within the walled City. He is distressed:

> though the Bill in general is abated, yet the City within the walls is encreased, and likely to continue so, and is close to our house there. My meeting dead corpses of the plague, carried to be buried close to me at noon-day through the City in Fanchurch-street. To see a person sick of the sores, carried close by me by Gracechurch in a hackney-coach. My finding the Angell tavern, at the lower-end of Tower-hill, shut up, and more than that, the alehouse at the Tower-stairs, and more than that, the person was then dying of the plague when I was last there, a little while ago, at night, to write a short letter there. . . . To hear that poor Payne, my waiter, hath buried a child, and is dying himself. To hear that a labourer I sent but the other day to Dagenhams, to know how they did there, is dead of the plague; and that one of my own watermen, that carried me daily, fell sick as soon as he had landed me on Friday morning last . . . and is now dead of the plague.[36]

And again:

> it troubled me to pass by Coome farme where about twenty-one people have died of the plague, and three or four days since I saw a

> dead corps in a coffin lie in the Close unburied, and a watch is constantly kept there night and day to keep the people in, the plague making us cruel, as doggs, one to another.[37]

Cruel as dogs people were, particularly those who oversaw the parishes, where plague victims, dying in the streets, were sometimes chased or removed to somewhere outside their parish, so that the parish would not have to pay for their burial, however meager an expense that was. In Dorset, where the plague was brought by goods in a peddler's pack, no one would nurse the sick. The local magistrates turned to a young woman who had been sentenced to death for some crime or other, and offered to obtain a pardon for her if she nursed the sick at the local pesthouse. She agreed, performed her duties, escaped infection—and then, after the outbreak abated, was hanged nevertheless.[38]

Pepys remarks upon "the madness of the people of the town," who follow the death carts in droves to the edge of the plague pits to watch the burials. He also tells the story of a child taken by stealth from one of the boarded-up plague houses:

> Alderman Hooker told us it was the child of a very able citizen in Gracious Street, a saddler, who had buried all the rest of his children of the plague, and himself and wife now being shut up and in despair of escaping, did desire only to save the life of this little child; and so prevailed to have it received stark-naked into the arms of a friend, who brought it (having put it into new fresh clothes) to Greenwich; where upon hearing the story, we did agree it should be permitted to be received and kept in the towne.[39]

In poor neighborhoods, there were the bells, tolling at all hours, the rattling of the carts at night, the tears of mourners, the cries of the dying. Sometimes plague patients could be seen running down the streets, shrieking from pain or madness; other times the cursing and swearing of the corpse bearers were heard as they dragged the dead bodies from their homes by grappling hooks, tossed them into the death carts, and whipped their horses to prod them onward.

Sometimes the corpse-bearers did not stop to make absolutely sure their victims were dead before dragging them off with their hooks. Daniel Defoe recounts one apparently factual story in his famous novel.[40] A drunken piper passed out in a London street, and a cart came along and slung him on top of that night's harvest of the dead. The jouncing of the cart must have awakened him; he sat bolt upright, pulled out his pipes, and began to play. This so terrified the bearers that they abandoned the cart in the street and fled, insisting afterward that they had picked up the devil himself in human form.

The shutting up of plague houses, ironically, probably did more to spread the infection than to contain it. It killed many that might have lived, through being shut up with the dying and their fleas; furthermore, the nurses who attended the dying, purloining what they could of goods and of clothing, inevitably spread the infection as they moved from house to house. The seasonality of the plague, its rapid explosion during the summer months, and its erratic movement across the city, strongly suggest that rats and *Xenopsylla cheopis* were involved in the epidemic's overall pattern.[41] But within the homes, *Pulex* played a major part, and in the transfer of infected goods from house to house the infection inevitably spread.

Blanc and Baltazard, the mid-twentieth-century French plague experts who worked in Kurdistan and Morocco, and who more than anyone else focused on the role of *Pulex* in the spread of human plague, argue that rat plague is what moves the disease along, slowly and irregularly, but that it is the human fleas—*Pulex irritans*, in the case of Western Europe and elsewhere—that cause sudden intense and murderous explosions within households. The disease is spread by the manipulation of the corpse, the clothing of the deceased, the bedding—in short, by anything that conveys infected fleas from one household to another.[42] Bell relates how the infection entered the Isle of Wight through a single traveler: the mayor of Newport ordered the man's house shut up, but two women, changing and airing the dead man's sheets, came down with fulminant virulent plague and died. No one would touch their corpses, and they had to be buried in the householder's garden.[43]

The infection in London had to run its course; by October the Bills of Mortality began dropping, but the disease would not vanish altogether for another year. The reported toll was over 97,000, and some claim that the actual total was closer to 110,000.[44] The vast majority of plague cases fell in the poorest parishes outside the City: the very poorest parishes, the twelve in the outskirts of London, reported 21,420 plague deaths, 75 percent of all deaths.[45]

The plague faded out, and London returned to life; the court returned, and people picked up the threads of their existence that had been so badly frayed by the trauma of the year before. Nothing had been learned, or very little, and nothing the municipality or the parishes had done to abort the epidemic had done anything other than prolong it, intensify it, and increase the people's misery. The story was repeated, on a tiny scale, in the small town of Eyam in the following year; on learning that plague had entered the village in a box of tailor's samples, the virtuous rector of Eyam refused to flee the village and abandon his flock, though wisely he sent his children away. He convinced the entire village to shut itself up and remain in place; he argued that only by shunning contact with neighboring villages could the people of Eyam prevent the infection from spreading. Virtuous he was, but misguided: keeping the people at home among their fleas meant that 267 people, eight out of every nine of the inhabitants, died, including the rector's faithful wife, Catherine, who would not leave her heroic husband.

Perhaps the only worthwhile lesson of the plague of 1665–1666 was that of Scotland's successful attempt at a plague blockade. The Scots imposed a complete interdiction of trade from England, and all travelers were made to endure an entire forty-day quarantine in country houses set aside for the purpose before they were allowed to cross the border into Scotland. Plague was easier to prevent than it was to cure, as the Scots learned, and the entire country remained plague-free.[46]

Although no one could know it at the time, the Great Plague was the last wave of epidemic plague in England. Indeed, at this time, plague began slinking away from Western Europe altogether, fading back into the Eastern world. No one knows exactly why, though quite possibly there is a simple answer, as evidence from the Continent suggests.

* * *

If England during the Great Plague was among the most backward countries in Europe for public health measures, Italy was the most progressive. Shortly after the Black Death broke out in northern Italy, as early as 1348, the first hesitant steps were taken toward the creation of a genuine public health policy. By the early seventeenth century, this process was far advanced: the city-states of Milan, Genoa, Florence, and Venice had permanent boards (magistracies) of public health. Even small cities and rural communities had their own temporary health boards in time of plague, which were under the direct command of the permanent boards of their respective capitals.[47]

The first, and perhaps most onerous and expensive, task facing these magistracies was to keep plague out of the region. Travelers from outside the region had to be examined for evidence of plague: this took medical workers with some knowledge and experience, who had to be employed on a full-time basis. Heralds had to be kept on call, as it were, to carry messages from one city to another. Records had to be kept, expeditions undertaken to purported plague regions. These tasks required a large and expensive bureaucracy—and all of that had to exist above and beyond any emergencies dictated by a actual outbreak of plague.[48]

The Italian city-states did not rely on ignorant old women to certify who did and did not die of plague. They hired medical and public health workers to make those determinations. In the event of plague, families were placed in quarantine. Sometimes this involved being immured in their homes, as families were in England; there were also pesthouses, some of them quite large, placed outside the city walls. Patients were confined within them, whether they wanted to be or not. In Genoa, the families and close contacts of the afflicted were placed in a separate section of the pesthouse for forty days, a policy that, however debilitating it may have been to the spirit, was far less dangerous to the body than being locked up directly and in close quarters with the sick and dying. After their quarantine, the surviving relatives were still confined for a further period known as "convalescence."[49] In the Genoese pesthouse, there was still another section where people "under suspicion" were confined. These were people who either suffered from unidentified fevers or who had come from areas where plague was suspected.

All of these plague suspects were ordered to strip and then carefully examined by the resident physician for any sign of the disease.

Furthermore, Italian city-states had widespread regulations—deeply resented by the public, who hated the governmental intrusiveness—about the burning of suspected plague goods.[50] The mattresses and clothing of those under suspicion were frequently burned, and new ones provided. In Florence, during the epidemic of 1630, inspectors sent by the board of health had the mattresses of even healthy people burned, on the grounds that they were "filthy and fetid"—a true preventive public health measure.[51] In Spain as well, according to a document found in the National Library of Madrid, during the outbreak of 1600–1601, the authorities took the same approach, keeping plague patients in strictest isolation, and burning clothes and bedding of the sick person "even if he should recover."[52]

Italian physicians had their own theories about why mattresses had to be burned and cloth shipments embargoed, and why streets needed to be cleared of stinking rubbish. Rubbish produced miasmas, and miasmas corrupted the air and gave rise to plague. Dirty mattresses had to be incinerated because they also stank, adding to the air's corruption.

Even the famous plague doctors' floor-length, hooded robe of thin waxed cloth and a long beak had been designed with the same principle in mind: it too was meant to repel miasmas. The filthy atoms of the plague were thought to catch on and cling to coarse fibers, on fur, wool, or any other cloth, which would infect the physician in the pesthouse. Fine thin cloth with an aromatic wax coating, perfectly slick and smooth, would prevent miasmal atoms from sticking, and therefore protect the physician from the pesthouse's corrupted air. The spices in the costume's beak would serve to purify the air the physician breathed. These robes were originally manufactured in Marseilles, but their use spread widely in Italy during the seventeenth century.

The robes seemed to have a protective effect, though not everyone had faith in them. Father Antero Maria da San Bonaventura, the able young administrator of the principal Genoese pesthouse, during a massive outbreak in 1657 complained that "the waxed robe in a pesthouse is good only to protect one from the fleas which cannot nest in it."[53] The pesthouse fleas were ubiquitous, tormenting. Father Antero describes his battle with them in the pesthouse:

> I have to change my clothes frequently if I do not want to be devoured by the fleas, armies of which nest in my gown, nor have I force enough to resist them, and I need great strength of mind to keep still at the altar. If I want to rest for an hour in bed, I have to

> use a sheet, otherwise the lice would feast on my flesh; they vie with the fleas—the latter suck, the former bite. Someone may remark, "Oh, what nonsense you are saying." Reader take pity on me, in the greatest of my sufferings. I can swear to you that all the bodily torments which are of necessity suffered in the lazaretti [pesthouses] cannot compare even to the fleas, for they do not leave me alone even in the coldest depths of winter.[54]

Father Antero would not have known it, of course, but he is telling us something critical in this passage: these fleas were *Pulex irritans.* They nest in his clothes and on his bed; they mass in swarming numbers, they do not leave him alone "even in the coldest depths of winter"—when *Xenopyslla cheopis* would be clinging to its rat for survival. Perhaps the dedicated father wore a waxed robe, perhaps he did not, but certainly in his own bed he was exposed to the murderous sorties of multitudes of *Pulex*, and he did contract a dangerous case of plague, though he recovered and went on to administer his pesthouse.[55]

Though the pesthouses served the admirable function of keeping plague patients and their infected fleas away from the populace at large, their conditions were frightful, less a modern hospital than an antechamber of hell. During the plague at Bologna in 1630, Cardinal Spada described the pesthouse:

> Here you see people lament, others cry, others strip themselves to the skin, others die, others become black and deformed, others lose their minds. Here you are overwhelmed by intolerable smells. Here you cannot walk but among corpses. Here you feel naught but the constant horror of death. This is the faithful replica of hell since here there is no order and only horror prevails.[56]

As a social institution, the pesthouse had little to offer over the heartless English practice of nailing the afflicted together with her family into their own home. But in other ways the Italian system was far advanced over the British: the burning of the mattresses and clothing of the dead, and replacing them at the town's expense, were excellent public health tactics. The Italian health magistrates anticipated the World Health Organization by several centuries, though the rest of Italy and Europe would not follow their example. The sophisticated attempts of the health magistracies of Florence and Genoa to set up an international health board came to nothing when other Italian cities refused to join them.

Still, the health regulations in Italy, mired as they were in a series of false assumptions, represented a notable advance in the development of public health. They also may have had a salutary effect on the plague epidemics themselves, though this is difficult to quantify. The death rate in Pistoia was remarkably low—less than 3.5 percent of the population.[57] In the Florentine outbreak of 1630, out of a population of 76,000, nine thousand died of plague—a formidable number, but a long way short of the Great Plague in London. In the later Italian outbreak of 1656–1657, strict regulations kept plague out of Florence altogether, while Naples, far more lax, lost nearly half its population of 300,000.[58]

But in 1656, health regulations did not save Genoa once the plague had entered. Forty-five thousand people were lost that year to the plague—60 percent of Genoa's population. As the Scottish, and now the Florentine, experience shows, it was far better to keep plague out of the area than to deal with it once it had entered the population.

Plague would not trouble Italy much longer; the 1656–1657 outbreak was the last march of the pestilence across that peninsula. Plague raged in England, Italy, and Spain, and then died down quickly. By the close of the seventeenth century, plague had vanished from Europe.

But then came Marseilles in 1720, and the worst epidemic of all since the Black Death.

* * *

Once again plague came from the Levant, Russia, and Asia. The harbor city of Marseilles, accustomed to receiving merchandise from the Middle Eastern ports on the Mediterranean, had instituted regular quarantine procedures for all arriving ships. The crews were made to stay in the pesthouse while the goods in the holds were unpacked, inspected, and aired. For ships thought to have been exposed to plague, there was a further regimen—the crew and merchandise were placed on Jarre Island, one of an archipelago of four islands off the Marseilles coast.

This regimen should have been enough to protect the city. But it was not. One small oversight, through denial and official inertia, ballooned into a catastrophe—all the more horrifying, in retrospect, because it was so easily preventable.

The *Grand St. Antoine*, a French merchant ship, had sailed from Sidon on the last day of January 1720 with a load of cloth on board. It stopped at Tripoli in Lebanon, where it picked up some Turkish passengers and more cargo. The captain of the ship did not learn until he came to Tripoli that plague was raging in Syria: by the time he put into Leghorn in Italy, one Turkish passenger and eleven of his crew were already dead. Under these circumstances, Captain Chataud assumed that the more stringent quarantine regimen would be put into place, and he naturally expected to be quarantined off Jarre Island. But they were told to proceed to the lower-security infirmary, where the disease was first misdiagnosed as a fever, but not plague. After buboes were recognized on some of the sailors, the rest were supposed to be removed to Jarre, along with all the goods, which were to be burned. But somehow things did not work out as they should have. The passengers taken to Jarre were assigned for some reason to a "light" quarantine of fifteen to twenty days, instead of the usual forty. Along with their clothing and personal effects, passengers began arriving in the city. Apparently some also sold some contraband cloth goods from the ship in the city, goods that should have been destroyed.[59] The plague began to spread.

The infirmary doctor, out of guilt, denial, or ineptitude, continued to deny that plague was abroad in the city, but the local doctors found themselves with clear evidence—people with buboes, people with pestilential fevers, people dead of plague. But officially the outbreak was kept quiet. As soon as they were diagnosed, the plague victims were brought to the infirmary: as they died, they were buried secretly at night on its grounds.

City officials did nothing. No one wanted to cordon off the city; they feared panic and an economic breakdown. On July 23, fourteen dead bodies were discovered on a single street.[60] By August, panic was rising, and the provincial parliament in Aix en Provence put the city under a general quarantine.[61] It was too late—over ten thousand people had already fled, scattering plague throughout Provence. Meanwhile, a halt to all commerce meant that people trapped in Marseilles had no flour for bread, and they began to riot. Grain markets on the roads to Aix and Toulon were eventually reestablished, but there was now no meat and no wine.

People continued to die. Mass graves had to be dug, and once again people were thrown in until the pits were filled. The streets and boulevards filled with the desperately ill, who had nowhere to go, and who fled down to the harbor and the stretch of trees on the Rue Dauphine, which ran down to the hospital. When the bodies were thrown into huge piles, those at the bottom split and burst, oozing infection and stench into the streets, like the death pits of Constantinople, or the mass graves in London.[62]

The city did not know what to do. They feared that, flung into the sea, the bodies would float in the harbor, polluting the water, and washing up against the shore. Some were stuffed into holes in the ancient city wall and covered with quicklime. Eventually more grave pits were dug, and the bodies removed from the streets.

It took two years for the fury of the plague to spend itself. By the time it was over, 80 percent of those infected had died, making this outbreak exceptionally lethal for bubonic plague, and suggesting that here, too, *Pulex irritans* and septicemic plague had been at work. Whether, or to what degree, rats helped spread the disease, we will never know. Nonetheless the plague appears to have come from direct contact with infected passengers and infected goods, from cloth that surely bore infected *Pulex* in its folds. The rapid spread of the infection, and the early sight of people dead in the streets, suggest a virulent plague, borne by human beings and human fleas.

It was the last flare of an old nemesis. With the close of the Marseilles epidemic, widespread plague disappeared forever from Western Europe. Except for a few sporadic rat-borne cases brought by ship in the early twentieth century, it has never returned.

* * *

One of the enduring mysteries of this most mysterious affliction is why, after flaming across Europe over and over again since the fourteenth century, with a change in its mode of transmission, but with no diminishment of its power, did plague disappear? Plague had been more or less endemic in England for centuries: in between major epidemics, an individual case here, a small outbreak there, were regular events, hardly noticed. But after Marseilles, the disease simply vanished, retreating to the Levant, the Middle East, Turkey, and Russia, where it broke out in two terrible waves in the mid-eighteenth century. Then it receded from those redouts, too, back into its ancient reservoirs in Central Asia, northern Central Africa, China.

Many theories have been advanced, and none proven. Some scholars have suggested that silent infection with one of the two related *Yersinia* species, *Yersinia pseudotuberculosis* and/or *Yersinia enterocolitica*, helped inoculate the European rat population against *pestis*. An article by French scholar J. M. Alonso puts it this way:

> Experimental infection by Y. enterocolitica, inducing a transitory and spontaneously cured infection in the immunocompetent host, only inducing opportunistic infections in the immunodeficient host, promotes efficient immunity against plague. Thus, it seems likely that the emergence of some variants of Yersinia, less virulent than Y. pestis, but able to induce a long-lasting protective immunity against plague, have contributed to its eradication by a silent enzootic infection among the wild reservoirs of rodents.[63]

This is intriguing, if unprovable. But plague in Europe has always been a disease of commensal (domestic), not of wild, rodents. Elsewhere in Eurasia active plague foci continue to exist—and pseudotuberculosis has done nothing, after several centuries, to eliminate them.[64] In the one region, therefore, where we might actually expect to see evidence of such replacement of one *Yersinia* infection by another, there is no such evidence at all.[65]

* * *

Did improved public health systems play a dominant role in keeping plague out of Western Europe? A strict cordon sanitaire was imposed along the boundaries of the Austro-Hungarian and the Ottoman empires, using the centuries-old techniques of quarantines, isolation, burning of suspect goods,[66] and even shooting of peasants with the plague or of anyone who tried to sneak across the borders.[67] A vivid picture of this border can be found in A.W. Kinglake's famous travel book *Eothen*, which was first published in 1844, and which describes the author's adventures in the plague-ridden and mysterious East:

> The two frontier towns are less than a gunshot apart, yet their people hold no communion. The Hungarian on the north, and the Turk and the Servian on the southern side of the Save, are as much asunder as though there were fifty broad provinces that lay in the

> path between them. Of the men that bustled around me in the streets of Semlin, there was not, perhaps, one who had ever gone down to look upon the stranger race dwelling under the walls of that opposite castle. It is the plague, and dread of the plague, that divide the one people from the other. All coming and going stands forbidden by the terrors of the yellow flag. If you dare to break the laws of the quarantine, you will be tried with military haste; the court will scream out your sentence to you from a tribunal some fifty yards off; the priest, instead of gently whispering to you the sweet hopes of religion, will console you at dueling distance, and after that you will find yourself carefully shot, and carelessly buried in the grounds of the Lazaretto.[68]

Kinglake himself was trapped for nineteen days in Cairo, while severe plague raged around him:

> Once I was awakened in the night by the wail of death in the next house, and another time by a like howl from the house opposite; and there were two or three minutes, I recollect, when the howl seemed to be actually *running* along the street.[69]

The Europeans living in Cairo were consumed by fear of contagion, and kept themselves in the strictest quarantine: Kinglake, however, did not believe that the mere touch of a sleeve, or the opening of a letter, could pass on the disease, and he went on about his life in as normal a manner as he could, though for caution's sake he did try to avoid physical contact with people he passed in the streets.

He did not contract the plague, though most of his acquaintances did. But it was not individuals shutting themselves away from plague that brought the outbreak to an end. It was the strict health regulations imposed now within Turkey and Egypt that seem to have eventually driven plague from these areas. As the

British plague fighter and researcher W. J. Simpson put it in his 1905 work, *A Treatise on Plague:*

> The retrocession of the plague from Egypt and Turkey was so remarkable an event, and followed so closely on the organization of protective measures, being not more than 7 years in one case, and 14 years in the other, that it is difficult to dissociate from them the relationship of cause and effect.[70]

Still, Simpson argues that we ought not to exaggerate the benefits of sanitary regulations, as plague had been receding eastward since the seventeenth century until the middle of the nineteenth, when, in the course of five years, from 1839 to 1844, it disappeared entirely from its old haunts in southeastern Europe, the Levant, and Egypt.

So what happened? The French researchers Blanc and Baltazard have proposed a solution to the riddle that is so simple, so trivial, that it seems preposterous. But they may be right. It hinges upon "les pains de savon"—cakes of soap.

Blanc and Baltazard, who were for much of their careers literally voices in the wilderness, who spent their lives researching plague outbreaks in Kurdish Iran and in Morocco, insist that infected rats may originate a plague outbreak within a community, but they cannot maintain it. In the absence of human fleas, they maintain, there can be no intense explosion of bubonic plague (they are not speaking of pneumonic plague). They put it this way:

> One can affirm that, without human parasites, an epidemic of bubonic plague is not possible.
>
> It then becomes easy to understand that which took place in the epidemics in Europe in the fourteenth through the eighteenth centuries. The virus [sic] came with the rats (always *Rattus rattus*—the black rat), which sporadically touch man through the rat flea—

being almost exclusively in that epoch *Nosopsyllus* and *Leptopsylla.* The plague in man always found the most perfect conditions for spread, thanks to the intense parasitism in the populations of that era. The plague retired from Western Europe, like typhus in the eighteenth century, at the moment when cakes of soap appeared, when the density of parasites began to roll back in the face of hygiene. At the present time, epidemic human plague is found exclusively in those countries where human parasites are still dense, even though rat plague continues to be endemic in all countries.

Those countries where individual human hygiene is good, but which the habits of living still favor contact with man and rodents, still see relatively frequent cases of plague, but these plague cases never turn into epidemics.[71]

Most plague researchers consider plague a dead-end infection in humans. To them, it is fundamentally a zoonosis that drops into the human species when it cannot find any more rats to infect, or when a human being wanders into a natural focus and contracts it, as it were, by mistake. Inarguably that is the pattern of plague today. But plague has always had the capacity to become a human disease, whether one borne by human parasites, or directly, lung to lung, in its pneumonic form.

It is this capacity of plague to be a *human* disease that Blanc and Baltazard force us to recognize. It was not a change in rat species, nor the spread of another form of *Yersinia* into populations of native rodents, nor even the imposition of sanitary quarantines that forced the plague out of Europe and much of Asia. It was a change in *human hygiene* that drove plague out of humanity altogether, back to being a true animal disease again. It was the irresistible spread of inexpensive cakes of soap.

If Blanc and Baltazard are right, that part of humanity which remains free of human parasites has little to fear from naturally occurring epidemic bubonic plague. But as we will see in the next two chapters, that does not mean that plague itself is no longer a threat. We may not see a resurgence of the disease that tormented the people of Marseilles, of London, of Cairo. But we are still at risk—at more risk than ever—of plague in its most terrible form, plague that all the soap in the world cannot wash away.

# VI

## THE THIRD PANDEMIC

**But he will find that nothing dares**
**To be enduring, save where, south**
**Of hidden deserts, torn fire glares**
**On beauty with a rusted mouth,—**

**Where something dreadful and another**
**Look quietly upon each other.**

LOUISE BOGAN

Medical officers during plague outbreak, Harbin, 1921.

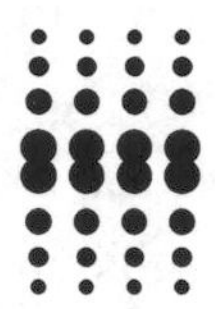

Plague, as it moved out of Europe, became less a human disease, and once more largely a disease of rodents; the remnants of the Second Pandemic drifted back over time ever closer to their point of origin, so long ago, in Central Asia. The last European outbreak occurred in Russia in 1877–1889, near the city of Astrakhan on the European side of the Ural Mountains, just north of the Caspian Sea. It was thirty-three years after the final episode in Turkey. Two hundred people were affected with "ambulatory" bubonic plague—painless swellings accompanied by a low fever—and only one died.[1]

But within a year a far deadlier form of plague had erupted, killing 420 people in the neighboring Cossack village of Vetlianka and surrounding areas. At the end of the outbreak, the Vetlianka plague proved fatal to everyone who contracted it.[2] Vetlianka, it turns out, is near the site of a very dangerous plague reservoir, the bobak, a type of small marmot.[3]

Vetlianka proved to be the last gasp of endemic plague in Europe. While India had suffered intermittent plague outbreaks since at least the seventeenth century, the late nineteenth century also proved a sort of respite. In Central Asia there were repeated small episodes, as we shall see; but there are always small episodes in Central Asia, from time immemorial down to the present day.

Toward the end of the nineteenth century, an entirely new strain of plague broke out in China. For some reason we don't yet understand, this rat plague variety, called *Orientalis*, has proved spectacularly successful, spreading through China, India, Southeast Asia, South Africa, Madagascar, the Western United States, and much of

South America. Through rats it has spread to all corners of the earth.[1] *Orientalis* plague, unlike marmot plague, has never become a true human disease: it has only rarely produced chains of pneumonic plague in human beings, and these do not last very long. The *Orientalis* strain, because of its ubiquity, has become the very model of the plague germ in the eyes of modern scientists. The model of rat and rat flea transmission was elucidated with the *Orientalis* strain, the intimate linkage between rat die-offs and human epidemics was demonstrated with *Orientalis*. It is *Orientalis*, brought by ship from China in 1900, that slaughters prairie dogs and ground squirrels in the Western United States. Everything we think we know about plague as a rodent disease that occasionally infects humans, we really know about the *Orientalis* strain.

When plague researchers speak of the Third Pandemic, that great oceanic wave of rat-borne plague that spread across the world, it is the *Orientalis* strain they generally mean. But at the same time that plague fighters and researchers in Accra, Yunnan, and much of India tried at once to contain and to understand this massive spill of rat-borne plague, another separate, different endemic was ravaging another, smaller corner of the earth. Marmot plague, which is never long absent from Central Asia, burst out in great chains of human-to-human pneumonic transmission in 1910–1911, 1917, and 1920–1921. Nearly 100,000 people died in these three epidemics, which had nothing to do, clinically or epidemiologically, with the outbreaks of rat-borne plague so widely dispersed now throughout the world. The so-called Third Pandemic, therefore, was really two distinct epidemic processes, related only by time.

In a sense, the Third Pandemic gives us an opportunity to examine the Black Death and the Justinian Plague writ small and close at hand. The First Pandemic was predominately rat-borne plague; early-twentieth-century rat-borne plague in India and China gives us a pale idea of what the Justinian Plague might have looked like on the ground. By examining the patterns and effects of rat plague in China and India, we have some indication of how plague ravaged the ancient world. At the same time, the far deadlier, if short-lived, explosions of pneumonic marmot plague in Manchuria and Shansi give us a faint sense of the horror of the Black Death as it crossed the earth and killed its multitudes.

* * *

The story of the Third Pandemic is also the story of the doctors and researchers who began to solve the mysteries of plague. This all happened very quickly. In 1894, plague broke out in Hong Kong; only months later, two separate plague researchers had seen the bacillus. Three years later, the connection between plague and rats had been made; not long after, P-L. Simond began to elucidate the role of the rat flea. In 1910, the young and heroic Dr. Wu Lien-teh helped to break the back of the first Manchurian pneumonic plague epidemic, and lay the framework for much that we now know about the disease in that form. No one alive has seen what these men saw: the great rat-borne epidemics of China and India, the fatal pneumonic plague in Central Asia. The accounts of these witnesses are as close as we can come to a scientific understanding of epidemic plague.

The history of plague in the Third Pandemic—with the exception of the less well-known Manchurian outbreaks—has been well documented.[5] The descriptions of its victims differ little from those of Boccaccio's day, or Justinian's. What is new are the radical, and unprecedented, discoveries these scientists made about the nature of plague transmission and of the bacillus itself. Though the chief discoveries were made in China and in India, rat plague is hardly the whole story. Marmots, too, finally come into the spotlight.

Rather than speaking of the Third Pandemic, then, we should speak of rat plague and marmot plague—the rat-borne plague of biovar *Orientalis*, which successfully colonized much of the earth, and the three short-lived but devastating marmot plague outbreaks of Manchuria and of Shansi, China.

The *Orientalis* rat plague seems to have begun in Yunnan, in the far southwest of China, at some point in the eighteenth century. That rats were connected to this outbreak from its onset there can be no doubt. A scholar writing in the late eighteenth century noted that

> In 1792 at Chaochow, Yunnan, rats were seen in daytime. They vomited some blood and fell dead. Human beings inhaling the odour of the dead rats rapidly succumbed. In 1792, Shi Tao-nan, a native of Chao-chow . . . a young man of extraordinary talent—composed a poem, *Tien Yu Chi*, a part of which entitled "Death of Rats" thus vividly describes the calamity:
>
> Dead rats in the east,
> Dead rats in the west!
>
> As if they were tigers,
> Indeed are the people scared.
>
> Few days following the death of the rats,
> Men pass away like falling walls! . . .
>
> The coming of the devil of plague
> Suddenly makes the lamp dim,
>
> Then it is blown out,
> Leaving man, ghost and corpse in the dark room.[6]

The "young man of extraordinary talent" died of plague only days after this poem was written.

Later reports also clearly indicate that observers in China made the connection between dead people and dead rats—indeed, it seems hard to miss.

There were numerous other outbreaks of rat-and-human plague in Yunnan, including one that followed the Taiping Rebellion (1851–1864). Wu quotes one Yu Yueh (1823–1908):

> In the early days of the reign of T'ung Chih, Yunnan was thrown into a confused state by the Taiping rebellion, and this was followed by a plague epidemic. Immediately preceding the epidemic, there was found a great number of dead rats in walls and ceilings. Human beings breathing the odour of putrefaction from the dead rats invariably fell sick. The disease began suddenly with the appearance of hard nodes in different parts of the body, which were slightly red in colour and somewhat tender, followed quickly by fever and delirium. . . . When the disease appeared in one family the surrounding neighbors immediately moved away, but in spite of that most of them died on the way.[7]

After that last outbreak, Yunnan was plague-free for three decades. Wu insists that the disease was never truly endemic in Yunnan, and that "sparks of infection" had come overland via caravans from Central Asia.[8] Other scholars maintain that plague was indeed endemic in Yunnan, in some of the ten species of native rodents, including, most importantly, the yellow-chested rat.[9] In any event, plague passed in an easterly direction, to Pakhoi, a port in southern Kwangtung; sporadic cases were seen there beginning in 1867, with major outbreaks in 1877 and again in 1882. There are a thousand difficult miles between Yunnan and Pakhoi, but, though tin and opium were the principal commodities brought from Yunnan to Pakhoi, somehow, and probably overland, the disease had spread.[10] By 1892, "a large number of people" died of plague a hundred miles east of Pakhoi, and still the disease crept eastward, to Canton and the world port of Hong Kong.

By 1894 the disease had reached Canton, killing about seventy thousand, in a population of 1.5 million.[11] Hong Kong was invaded by plague soon after, and despite heroic sanitary measures, the outbreak was as deadly and long-lasting as that in Canton, which took no sanitary measures at all. But despite the old understanding in China that rat-falls preceded human infection, systematic rat killing did not begin until 1901: many of the Western experts called in to assist with the epidemic did not take the connection seriously.

But Alexandre Yersin did. This strange, austere man was perhaps the first to understand how rats and plague were intimately related, and that rats formed a necessary conduit in bringing the disease to men. Alexandre Yersin was born in Aubonne, Switzerland, in 1863; he was an obsessive naturalist and observer from an early age, trapping and killing his mother's cat so that he could dissect it.[12] Shy, almost reclusive, he studied pathology so he could avoid contact with patients and doctors.[13] After studying tuberculosis for three years with the famous German scientist Robert Koch, he moved to Paris, where he found work as an assistant pathologist. In 1886, he accidentally cut his hand while autopsying a patient who had died of rabies. Fortunately, this occurred after Louis Pasteur had already developed his rabies vaccine; through his subsequent treatment, Yersin met both Pasteur and his assistant Emile Roux. Yersin went to work with Roux on the tuberculosis bacillus.

He stayed three years. But one day while Roux was absent from the laboratory, Yersin dropped off a long letter with instructions to his assistants on how to complete his experiments, and went to Southeast Asia as a ship's physician on a merchant marine line. Once in Asia, he moved from ship to ship, exploring and mapping the land around him whenever he was able.

The roving life of an itinerant geographer seemed to suit him; he did not mind privation, and he had a wiry frame with exceptional powers of endurance. It was not until 1894 and the outbreak of plague in Hong Kong that he ceased his wanderings in the jungles and mountains of Indochina for a time and returned to the laboratory, determined to find and identify the plague bacillus. He set up a tiny laboratory equipped only with a microscope and an autoclave (for destroying dangerous materials). One of Yersin's two assistants absconded with all his funds.

At the same time, a well-funded Japanese research group, headed by the stocky, imperious Shibasaburo Kitasato, a student of Robert Koch's, had already established itself in Hong Kong. Kitasato's well-stocked laboratory at the Kennedy Town Hospital, Hong Kong's principal plague hospital, was a striking contrast to the pitiful straw hut Yersin was allowed to build on the hospital grounds. He was denied entrance to the Kennedy Town Hospital; he claimed in a letter to his mother that Kitasato's entourage had bribed the hospital staff to prevent him from having access to plague corpses. Yersin complained to the governor, though, and eventually was given access to the corpses he needed in order to perform autopsies and to try to find the plague germ.

Despite all these disadvantages, it was Yersin, the superior scientist, who actually found and identified the plague bacillus; Kitasato found *something*, but his description was poor—he described his discovery as slightly motile, while the plague bacillus does not move or wriggle at all. He also incorrectly seemed to think it could be stained using Gram's method—a method of staining bacteria on a slide that reveals the presence or absence of certain cell wall components. Allowing bacteria to be classified broadly as Gram-positive or Gram-negative has significance for treatments, as anthrax is a Gram-positive germ, plague gram-negative. (It may be that Kitasato's samples were contaminated with pneumococci, the short, rounded bacteria that cause bacterial pneumonia.)[14]

Yersin took his plague samples back to Paris, to Roux's laboratory. Roux confirmed Yersin's description and suggested that Yersin work to develop a plague antiserum as a plague treatment. Yersin sailed once again for Indochina, establishing a branch of the Pasteur Institute in Nha Trang: he attempted to produce a plague horse serum in Vietnam, but it did not work very well. There were many such efforts, but none of them seemed to help much—at best, they prolonged the illness and delayed death by a few days. Until streptomycin was introduced in the 1940s, even the great physicians of the Third Pandemic were scarcely better able to treat plague than Gui de Chauliac.

But Yersin's discoveries were not limited to discovering the bacillus; in his classic paper of 1894, he clearly describes the connection between plague and rats:

> In the infected boroughs, many dead rats are found on the ground. It is interesting to note that in the part of the town where the epidemic started, a new sewage duct has just been installed. The pipes are much too narrow and difficult to clean. They constitute a permanent focus of infection. . . .
>
> The lodging quarters of the poor Chinese are often so revoltingly dirty that one scarcely has the courage to enter them. In addition, the number of occupants is unbelievable. Many of these slums do not have windows and are below the level of the ground. One can imagine the havoc that can be caused by an epidemic on such a terrain and the difficulty involved in its control. The only remedy was to burn down the Chinese town. This was proposed but was unfeasible for budgetary reasons.
>
> Few Europeans have been struck by the disease, thanks to the better conditions under which they live. European houses are not fully safe, however, since even there one can encounter dead rats.[15]

Later on in the paper, Yersin notes that "passages in guinea pigs are easily achieved with macerated spleen or blood. Death is more rapid after a few passages." Yersin was the first to discover that the virulence of plague might be increased by repeated passage through animals.

Yersin never returned to Europe permanently; he seems to have married a Montagnard woman and lived with her in Nha Trang.[16] But they had no children. Yersin died in 1943, still in Vietnam, and was buried on his riverside plantation as he had wished. His home has been preserved as a museum by the Vietnamese people.

Yersin believed that insects could carry plague; he ground up the legs and heads of dead flies he had found in his laboratory and injected the results into guinea pigs, which later died of plague. "The inoculum," Yersin says, "contained a large number of bacilli which were similar to the plague organism and the guinea pigs died in 48 hours with the specific lesions of the disease."[17] But it was left to the French researcher Paul-Louis Simond to discover plague transmission through rat fleas.

Though Yersin shunned any personal acclaim or attention, he nevertheless received the recognition he deserved for his discovery of plague, not least in the name of the germ he had identified, *Yersinia pestis.* But Simond's research, as lonely and thankless as Yersin's had been in his miserable straw hut, would confront a tidal wave of hostility. If Yersin was reclusive, Simond was nearly invisible; to this day, little is known about him. He may have been a missionary.[18] In a photograph Simond is dressed in a military uniform: he wears a medal at his neck and two on his breast pocket.[19] He has an erect bearing, a thick head of slightly greying dark hair, a bristling mustache, a small, neatly clipped beard, and an alert, direct gaze. He came to Bombay to help combat the raging plague epidemic of 1897. The prevailing theory of the Indian plague researchers of the time was that rats caught plague by cannibalizing their dead fellows, and that people caught plague through tiny cuts or cracks on their bare feet. Simond was unimpressed: he showed that it was actually rather difficult to convey plague to rats through feeding them. Furthermore, mere contact with infectious materials did not seem to infect rats at all. Pricking the feet of rats with a plague-contaminated needle infected them easily, while rubbing plague materials on the surface of an intact rat foot produced nothing. Simond pointed out that no one had ever shown that plague patients had sores on their feet.[20]

But the ability of tiny pinpricks to cause infection in rats made Simond wonder: could a sucking insect contaminated with plague produce the same effect? He found that a small percentage of plague patients had a tiny ulcer or blister, called a phlyctenule, usually on the lower leg, which contained a pure culture of plague germs in the fluid. A bubo always appeared nearby: if the lower leg was involved, the bubo might appear in the groin on that side, or behind the knee. Curiously, these sites of flea bite (for so the phlyctenules proved to be) were often a harbinger of returning health; they never showed up in rapidly progressive, fatal plague.

Simond reasoned from the appearance of these blisters that they must have been caused by the introduction of plague into the body by an insect bite. But what insect? Simond realized that rats are often infested with fleas, and far more so if they are sick. He also showed that rat fleas would bite human beings. Though he never understood the mechanism of blockage, he nevertheless is responsible for the crucial discovery that fleas form the bridge between rats and men. His most critical experiment showed that healthy rats could not contract plague from sick ones *in the absence of fleas.*

But no one believed him. The British Indian Plague Research Commission, as British plague researcher & author L. Fabian Hirst puts it, "considered Simond's experimental evidence so weak 'as to be hardly deserving of consideration.' . . . In Hong Kong, Hunter and Simpson failed to infect rats with infected rat fleas and concluded that 'plague infected fleas are of no practical importance in regard to the spread of plague.'"[21] It was many years before Simond's discovery received its proper and general acknowledgment.[22]

Meanwhile, the Third Pandemic raged. From Hong Kong, ships spread infection all over the world: to the United States and much of Latin America, to Australia, to India, to South Africa. In Bombay, plague exploded. Over the next thirty years, 12.5 million people would die in India alone. Almost all of these cases were bubonic plague; only 2 percent or so went pneumonic. India, more than anywhere else, provided the crucible in which the modern understanding of plague was formed, thanks to the early work of the Plague Research Commission, and the work of Simond. The final conclusions of the commission were as follows:

1. Rat plague is flea-borne.
2. Rat plague cannot be transmitted continuously as an epizootic in the absence of rat fleas.[23]

Hirst's own research shows incontrovertibly the connection between rat epizootics and human outbreaks. He describes the 1905–1906 outbreak in Bombay:

> In general, the poorest quarters suffered most. The epizootic was most intense in Mandvi, the harbor section where plague first broke out. Here there was an average of 3.1 plague rats and 0.6 human deaths per building, and 14.4 human deaths per 1,000 of population. The human epidemic was most intense in the adjoining ward Dongri, with one human death per building and a human plague death rate of 16.4 per 1,000 and 1.9 plague rats per building. On the other hand, the districts with least human plague per building were likewise those where fewest plague rats were found.[24]

These numbers are a long way from the Black Death. The human deaths are not clustered; 0.6 deaths per building sounds sporadic, intermittent—the very model of zootic plague.[25] Though the Third Pandemic in India killed 12.5 million people, that was over a thirty-year period. The Black Death, in the most conservative estimates, killed 25 million people in a year and a half—an entirely different disease process.

In Bombay it started with the deaths of Norway or field rats: the incidence of Norway rat deaths peaked in the third week of February 1906. The black or house rats began to die a little later; their death rate reached its peak in the second week in March. Human mortality peaked the last week in April. Clearly, Norway rats brought in the infection and passed it to black rats, who in turn bequeathed it to the people they lived among. A neater illustration of the relationship of rat to human plague is difficult to imagine.[26]

After the discovery of the mechanism of blockage by London plague researchers A. W. Bacot and C. J. Martin in 1914,[27] the commission correctly argued that *Pulex irritans* was a very inefficient vector of plague, since it rarely became blocked, and *Xenopsylla cheopis*, the Oriental rat flea, a very effective one.[28] The Indian plague epidemics are the perfect model for tropical plague infections: massive rat die-offs, swarming Oriental rat fleas, scattered human infections with no evidence of contagion.[29] We do not need *Pulex* infestation to explain plague in India.

The rat-flea-human pattern so meticulously illustrated by the plague researchers in India has formed the pattern for our entire understanding of how plague works and spreads. But it is only part of the story. The Indian plague researchers tended to take this pattern for the whole, and, by and large, later researchers have followed them. Though there have been dissenting voices, the general understanding of plague is that it is essentially a rat- and rat-flea-borne disease that sometimes infects humans. There were indeed several minor outbreaks of pneumonic plague in various places in India and Southeast Asia, including Assam, eastern Bengal, Calcutta, Ceylon (Sri Lanka), Kashmir, and the Punjab. Some of these —Assam and Bengal—were areas with "few rats," while in the Punjab these outbreaks were associated with rat-borne bubonic plague. Still, none of these pneumonic outbreaks were very large: "the outbreaks were remarkably circumscribed, seldom spreading far beyond the family first affected. . . . Groups of infectious pneumonic cases are always liable to crop up in areas subject to bubonic plague; yet, save in Manchuria, *this type of the disease has displayed very limited powers of diffusion during the present pandemic.*"[30]

This is characteristic of rat-borne plague: it certainly has the capacity, on occasion, to go pneumonic, but never with the strength and explosive power of marmot plague.

Meanwhile, the pneumonic plague outbreak in Central Asia in 1910–1911 showed how devastating human-borne plague can be. At the time, this terrifying pneumonic outbreak was closely watched and widely publicized in the outside world. But by now, it has largely been forgotten, except among Russian, Mongolian, and Kazakh plague experts, who have reason to know better.

* * *

The two great Manchurian pneumonic epidemics of 1910–1911 and 1920–1921, and a third, scarcely reported but massive outbreak in Shansi, China, stand out against a pattern of endemic plague. The marmots of the region were known by the native Manchurians and Buryats to suffer periodically from a peculiar disease that struck mostly in the autumn, before the animals settled down in their burrows to hibernate through the long winters. In an 1856 expedition, an investigator named Radde collected some legends about marmots from the Gobi desert:

> The natives of that remote region hunted the tarabagan, though they believed that, ever since one of their ancestors incurred the wrath of their tribal god, each hunter would be condemned to live his next life as a tarabagan. There was one part of the flesh of the shot animal which must never be eaten, the lump of fatty tissue in the axilla containing the axillary lymphatic glands, for that was the remains of the dead hunter.[31]

Wu Lien-teh, the great Chinese plague fighter, reports that a healthy marmot has a cry "pu pa, pu pa," which supposedly sounds like the Manchu words for "no fear, no fear."[32] A silent marmot is a sick marmot, and hunters learned to avoid them. A sick marmot has, also, a staggering gait; it is slow-moving, and the red swelling in its shoulders, the soul of the dead hunter, may be visible. If, having caught the marmot, the hunter is in any doubt of its health, they prick its foot; if the blood is coagulated and does not flow, they toss the body to the dogs.[33]

Long tradition, therefore, kept native Manchurian and Mongolian hunters away from plague-stricken marmots. But the Cossacks and Chinese hunters who began to flood the region in the mid-nineteenth century, in search of ever more valuable marmot skins, paid no attention to the warnings of the natives. The three great pneumonic plague epidemics of the early twentieth century seem to have originated from such misadventures. But they occurred against a backdrop of constant small outbreaks, beginning in 1894.[34]

In one such instance, a Cossack settled in Abogaitui hunted and ate a marmot. He died after four days of illness, having passed the disease on to eight other people, all who apparently died of pneumonic plague. Two were autopsied and a positive diagnosis made. Another Cossack family, living in a particularly isolated area, shot and skinned marmots, "to amuse themselves." They were warned by older men in the area, but paid no attention. After skinning one marmot they decided not to eat it, and gave the body—which weighed about fifteen pounds—to a girl of thirteen to throw away. The barefoot girl dragged the bloody creature through the grass, depositing it in a field some distance away, and returned home along the same path. She came down with buboes and died, but everyone else in the household escaped infection.[35]

A similar account from 1889 also described a young Cossack girl from Soktui as the first victim—but she died of pneumonic plague, and so did everyone else in her household, as well as in the house of relatives who had taken in and washed the clothing of the dead.[36]

There are many such instances, which all seem to occur among the interlopers in the region. But what made the later explosive epidemics of Manchuria possible was a new railway system. In 1896, a Russian-Chinese agreement allowed for the building of a railway across Manchuria, the northeasternmost province of China, straight to Vladivostok. Manchuria was a huge, rich province, the size of France and Germany combined—and one of its chief riches lay in the skins of marmots, in great demand in the West as a kind of imitation sable. Some two million skins were processed yearly.[37] In 1898 China allowed Russia to extend this line south to the sea from a central point in the rail nexus—Harbin. Part of this area, and two connecting rail lines, were also controlled by Japan, making the political situation, when plague did break out, delicate and volatile.[38]

Meanwhile, perhaps ten thousand strong, hardy Chinese immigrants from Shantung, eager to hunt tarabagan, poured over the border; they, no more than the Russian hunters, knew or perhaps cared about the old Manchurian prohibitions on hunting slow, sick, stumbling marmots. Tough men, they thought nothing of spending days exposed on the frigid steppe, gnawing on meat dumplings washed down with tea, and waiting for the tarabagans.[39] There were also, at this time of year, thousands of migrant agricultural workers leaving Manchuria to avoid the bitter winters and to spend the Chinese New Year with their families.[40] These men, who came from Shantung, a province subject to the frequent devastating flooding of the Yellow River, found better agricultural employment in the rich soya fields of Manchuria, and so worked as seasonal laborers, returning home when the growing season had ended.[41]

"The railway has been an all-important factor," says one observer, who traces the great 1910 epidemic to the neighborhood of

Manchuria station in the beginning of October. It began spreading rapidly among the tarabagan hunters and laborers along the railway lines. From October 13 to the 26th the disease spread from station to station, with no attempt by the Russian government to curtail it.[42] After that date, the Russian government took charge as well as they could: twelve medical officers arrived at Harbin, and stiff measures to counter the infection were undertaken. They managed to keep the Russian New Town generally free of plague, but the adjoining Chinese settlement of Fuchiatien was another matter. Chinese hunters, traveling along the railway lines, would spend the night in overcrowded, filthy inns, many of which were below ground and windowless. Many people would share large, long brick platforms, called k'angs, which were heated from underneath by charcoal fires, and used for eating, dressing, sleeping. Better conditions for the propagation of pneumonic plague cannot be imagined: people with diseased lungs would cough and spit, the bloody sputum would land on the k'ang; the air would be charged with germs and moisture. The inns were veritable cauldrons of infection.

The local Chinese administration was simply not up to the task; fortunately, the councilor of the Foreign Ministry back in Beijing, Alfred Saoke Sze, was. He turned to a young man, a Malayan-born Cantonese with a Cambridge medical degree, who at the age of thirty had already served for three years as associate director of Tientsin Medical School. The story of the enigmatic Dr. Wu Lien-teh is inseparable from the story of the three twentieth-century pneumonic epidemics with which he was so deeply involved.

Sze, who had met Wu in Penang, knew what he was doing when he chose Dr. Wu to run the Chinese anti-plague effort. Wu had expertise as a bacteriologist, having trained for several years at the Liverpool laboratory of Major Ronald Ross. He had also worked as a physician in Penang for three years, until his persistent and fearless involvement with the anti-opium movement made him too many powerful enemies in Malaya. Wu was also a passionate advocate of Western medicine, and had little sympathy for the theories of traditional Chinese healing.[43] He once rescued a wealthy Malayan planter from a massive abscess, which traditional medicine, resorting to bindings of herbs and unguents, had allowed to fester until a pound of pus had to be removed from the man's back. In gratitude for saving him, the planter "forgave" Wu his struggle against the opium trade, in which the planter had a substantial share.[44]

After securing the right from the Imperial Palace to conduct autopsies—something normally forbidden by Chinese custom—Wu left at once for Harbin, accompanied only by a Chinese medical student. The student fortunately spoke Mandarin better than Wu did—Wu was more fluent in English than in Chinese.[45] In Harbin, they found an appalling state of affairs: tensions, terror, whispers about fevers, blood-spitting, rapid death, and bodies in the streets. The Chinese administrators, one of whom, languid, pale, and slovenly, was an obvious opium addict, were less than helpful. Still, Wu was able in short order to locate the body of a dead Chinese woman and autopsy her, confirming both bacteriologically and clinically that she had died of pneumonic plague.

There was no doubt, then, what they were up against. But stopping the epidemic was a daunting prospect, especially given the fragile relations among Japanese, Russian, and Chinese governments, who all controlled areas of Manchuria. Furthermore, China stood to lose a great deal by the shutoff of international trade were the epidemic not brought under rapid control. Meanwhile the disease drew closer and closer to Beijing.

Fortunately, there were Russian specialists with whom Wu quickly established a good working relationship. Ensconced in the local Russian plague hospital, which served the Russian-controlled Railway Area, Wu found Dr. Paul Haffkine, nephew of the Russian-Jewish plague researcher Dr. Waldemar Haffkine, who had gained fame in India for his anti-plague vaccine and his research. Paul Haffkine had great confidence in his uncle's vaccine—too great. To his alarm, Wu found that at Haffkine's plague hospital no one wore masks, and when Haffkine invited him to visit a plague ward, he did not know what to do. Not wishing to insult his host, who told him cheerfully that all the doctors and attendants had been vaccinated, Wu slipped into the ward, keeping his face averted from the patients as he examined them, and hoped for the best. Haffkine smiled at his visitor's evident nervousness—but his confidence in his uncle's vaccine was mistaken. Many of Haffkine's staff were to die of plague.

The Chinese sector, Fuchiatien, was far more crowded and dirty than the Railway Area, and it had no such hospital. There were few doctors. Wu needed all possible assistance, and welcomed the arrival of the French plague expert Dr. Mesny, head professor from the Peiyang Medical College in Tientsin. The forty-three-year-old Mesny, having worked in a small bubonic plague outbreak in Tongshan, China, was convinced from his own experience and from the work of the Indian Plague Research Commission that plague was rat-borne, not spread directly from lung to lung, and not contagious. He did not want to listen to Wu's talk of protective gauze masks for staff and doctors, of isolation of patients and contacts, and of all the necessary steps needed to contain a rapidly spreading fatal disease. He told Wu that "his personal opinion was more reliable than that of a mere novice . . . and he was determined that the Chinese government accepted his more mature advice." In Wu's account, in which he refers to himself in the third person:

> Dr. Wu was seated in a large padded armchair, trying to smile away their differences. The Frenchman was excited, and kept on walking to and fro in the heated room. Suddenly, unable to contain himself any longer, he faced Dr. Wu, raised both his arms in a threatening manner, and with bulging eyes cried out "You, you Chinaman, how dare you laugh at me and contradict your superior?"[46]

Wu responded that he regretted that a talk he had intended to be a friendly one had resulted in such unpleasantness, and that, furthermore, he had no choice but to report the matter to councilor Alfred Sze in Beijing. He composed a telegram offering his resignation—but Sze would have none of it, and suspended Dr. Mesny instead.

The conflict had a tragic outcome: Mesny, upon receiving word of his suspension, dashed to Haffkine's hospital and, with no protective mask, began examining plague patients. Six days later he was dead of pneumonic plague.

Mesny's death had a galvanizing effect. The staff at Haffkine's hospital began to wear Wu's gauze and cotton masks, though they did not always wear them correctly, and some plague workers still lost their lives. But the masks became common, even in the streets. Wu became chairman of the entire plague control effort. Doctors and staff began streaming into the area from many different institutions, from China, Russia, the United States, and Europe. Wu divided Fuchiatien, the Chinese section of Harbin, into four sections and instituted house-to-house inspections. Plague patients and their contacts were sent to isolation sheds and railway wagons lent by the Russian Railway. The contacts and the sick were housed separately, and the contacts carefully watched. None of the sick were saved, but many potential new cases, perhaps thousands, were prevented by these protective means. Regulations, and the movement of people, were strictly controlled by the Chinese infantry. A special police force of six hundred was organized and placed under the direct authority of the staff doctors.[47]

But there still remained the issue of the dead. This was January in Harbin, and family members, to avoid isolation, often threw the corpses of their relatives into the streets, where the bodies immediately froze. Some of the sick, driven out of their homes by relatives, and told to get as far away as possible, died upon the streets, and froze where they lay, or sat, or huddled for warmth: a grotesque tableau.[48]

The sight of these bodies cast a further chill of terror on the town. As in the Middle Ages and the Renaissance, carts hauled off the dead to a burial ground beyond the settlement. But burial was impossible in the frozen ground. So there the bodies lay, row upon row of them. Some were placed in thin wooden coffins, most of which were not even nailed shut, so that arms and legs and heads protruded. Others lay exposed, contorted, on the earth.

Plague remains alive in frozen tissue, and Wu, visiting the burial ground with its long rows of frozen, posturing dead, knew that the corpses were a serious physical and psychological threat to public health. Burning, of coffins and corpses alike, was the only answer. But Chinese tradition forbade cremation—the town might have exploded in outrage. Wu had only one alternative, to apply for an imperial edict to permit mass cremation.

He got his written edict from the emperor, and the burning began. Tiers of one hundred bodies and coffins each were stacked like firewood. Kerosene was poured on the piles and set alight. The remains were swept up the next day and again burned, and the ashes buried in a common grave in the softened earth.

All the plague victims henceforth were sent immediately to the pits for cremation. And at that point, Wu tells us, the tide began to turn.

Little by little the death rate dropped off, not because of any treatment, but because fewer people became infected under the new regimen. Whenever possible, Wu preferred to keep contacts in the open air, where the disease was less likely to spread.

Harbin-Fuchiatien was the epicenter of the infection, but the disease had spread throughout Manchuria along the rail lines. When it subsided, some sixty thousand people had lost their lives. We can only wonder how much worse the epidemic would have been without intervention—another Black Death indeed, on the march to Beijing and beyond.[49]

The plague control authorities encountered stiff resistance from some quarters. A French Roman Catholic mission station run by two French priests and housing some three hundred people was instructed to report any plague cases. Claiming extraterritoriality, the mission treated sick residents secretly, and hid the mounting number of plague deaths, burying its people in secret, until after one month 243 had died; victims, says Wu, of "religious fanaticism."[50]

Many uneducated Chinese also resisted the plague control efforts: they saw that no one taken to the plague hospital ever returned. But in the Mukden area some of the villagers seemed to take the government's admonitions seriously. Dr. Dugald Christie, an English scientist who lived and worked in the area for thirty years, reported that a number of villages effectively quarantined themselves, allowing no visitors, and deputizing certain trustworthy men to act as delegates for the entire village: they would go to markets, obtain what was needed quickly, and return straightaway home, avoiding the plague-ridden inns. These villages remained plague-free throughout the epidemic.[51]

Plague experts learned a great deal from the horrors of the Manchurian epidemic. Suspicion immediately fell upon the tarabagan as plague carrier, though at first, on numerous expeditions, no infected marmots were found.[52] In a 1917 paper, Wu writes that in an earlier publication he had "provided ample evidence to show that under normal conditions plague did not exist among the marmots." The First International Plague Conference, chaired by Wu, issued a series of reports that are still valuable today. But Wu acknowledges that it was another scientist, the American plague researcher Richard Pearson Strong, who discovered the connection between plague and marmots: "It was Strong, working in Mukden, who first performed experiments of plague inhalation among tarabagan and demonstrated . . . that they could take pneumonic plague if exposed to the organisms sprayed in droplet form."[53] Wu later identified marmots as the original reservoir of plague.[54] He and other scientists were able to show that not only do marmots suffer from bubonic flea-borne plague—they have their own fleas—but that they are able to catch and transmit pneumonic plague both experimentally and in the wild—the only creature other than human beings able to do so.

It is this critical factor that may make marmot plague so much more deadly than other strains: according to Drs. Suleimenov and Atshabar of Kazakhstan, among other scientists from the former Soviet Union, marmot plague has developed a *pneumotropism*—a tendency, whether it is transmitted through a flea bite, through skinning an infected animal, or pneumonically, to make right for the lungs. Further, Suleimenov and Atshabar point out that data coming out of China in 1991 show this clearly: in marmot regions, 45 percent of human contacts developed pneumonic plague. A Chinese research paper published in 1993 states that "of the 391 human cases [between 1958 and 1992 in Qinghai Province], bubonic and septicaemic plague were caused predominantly through skinning and mealing [sic] of infected marmota or wild animals and liable to change to secondary pneumonia, which covered 47.67% of the total. Pneumonic epidemics spread directly from person to person via respiratory tracts occupied 47.05%, while the infection through rodents to fleas and to person [sic] accounted only for 2.31%."[55] In Vietnam, where plague is carried by rats, 98 percent of people developed bubonic plague. Furthermore, marmots injected with plague bacteria in their rear quarters are all coughing two or three days later. "You see this with susliks as well—but never with gerbils or rats," say Atshabar and Suleimenov.[56]

And lastly, though this was not recorded by the researchers of the Third Pandemic, Kazakh scientists have found marmots dead in the field, with bloody saliva on their muzzles.

Human fleas seem to have played no role in the Manchurian outbreak. As one physician puts it:

> despite of the crowded state of the native inns during the return of the coolies to Shantung, dark and ill-ventilated chambers in which men lie huddled on the k'angs like herrings in a barrel, and despite the fact that during the rigors of the Manchurian winter the coolie neither washes nor even removes his heavy clothing, fleas are apparently uncommon in the winter season in this region. I personally never observed a flea on any of the patients.[57]

The same physician stopped at an old cigarette factory in Chi-fu that the migrant workers used for lodging. They found six workers sitting on a small k'ang about four feet square.

> They made no complaint and did not look particularly ill, but hearing one of them cough, I suggested that we should make him spit into a piece of paper. He was spitting blood. We made the others spit also. They were all spitting blood. The whole six were dead by the evening of the following day.[58]

A series of important experiments by American researchers Richard Strong and Oscar Teague, who worked at Mukden, have shown how easily pneumonic plague can be spread from lung to lung via droplet infection. Some of these large droplets could be seen, but many were so small as to be invisible. Strong and Teague set up petri dishes at certain distances from the mouths of patients, and measured whether plague colonies would form on those dishes. They learned that plague could spread at a distance of about six feet, that it was propelled by coughing or sneezing, but not much by ordinary talking, and that the floor easily became infectious from sputum. Bedding, too, became contaminated—a series of experiments carried out by Wu, Chun, and Pollitzer during the 1920–1921 Manchurian plague showed that plague in sputum—even dried sputum—could last up to eight hours in direct sunlight, which normally kills plague bacteria very quickly. Plague sputum on wool or gauze could remain infectious as long as seventy-two hours out of the sun, and for seven hours directly upon soil placed indoors. Covered by earth, plague sputum remained infectious for twelve hours, as long as the experiment ran; apparently, sputum mixed with earth survived for many hours as well. Liquid disinfection with a number of agents, including carbolic acid and lysol, had no apparent effect on plague sputum, though alcohol and lime seemed to kill the bacteria. Plague-infected clothing also produced positive plague colonies.[59]

All of this work confirms that pneumonic plague is a highly contagious disease and that plague germs, mixed with sputum, are a surprisingly resilient agent. Contamination of surfaces is a serious threat, and plague can easily be conveyed from lung to lung by coughing and sneezing. Pneumonic plague hospital wards, in the days before antibiotics, were death zones; and many people who did not wear masks, or who wore them improperly, paid, like the intemperate Dr. Mesny, with their lives.

* * *

By 1918, Dr. Wu, who had now immersed himself in plague prevention, had helped establish the North Manchurian Plague Prevention Service, a Chinese parallel to the great anti-plague system scattered like an archipelago across the face of the Soviet Union. The lessons learned in 1910–1911 had not been forgotten, but the outbreak of another epidemic, this time at Shansi and Inner Mongolia in China, showed how difficult the implementation of those lessons could be. This outbreak has been scarcely recognized and reported even to this day.

By 1917, the Chinese emperor had been overthrown, and the short-lived republican government of Sun Yat-sen installed instead. This administration lacked the force to handle the epidemic. As Wu puts it, "Although the then Republican Government displayed much interest in and concern over the epidemic, its several mandates and orders did not appear to possess the sting or power of the edicts issued by the imperial throne. There were too many authorities sent to the scene, each with its own head, opinion and plan of campaign to deal with the threatening situation."[60]

The outbreak began in late November 1917, when sporadic cases around Patsebolong, Inner Mongolia, near the Yellow River, evidently introduced a panic: fleeing plague victims passed through the important market center of Paotow, and beyond that town to other towns, including historic Tatungfu, where stand huge carved Buddhist statues dating from the first century C.E. Traveling by carts, on foot, horse, or donkey, hundreds of people fled the plague centers, carrying the infection with them. They came to Fengchen, a town of about nine thousand, and the end of the Beijing–Kalgan railway line, a month after the exodus began.

Dr. Wu reached Fengchen on January 3, to find two American plague workers, Drs. Frank Lewis and Ekfelt,[61] already in place. They had found that in the large town of Kueihua, a city of 200,000 to 300,000 already threatened with plague, they were not permitted to work: the local governor, says Wu, "denied that such a thing as plague ever existed, would not permit the doctors to examine any cases and refused to stop any eastbound traffic from the infected area of Paotow."[62] These American doctors joined Wu's staff: they tried to implement several proposed regulations, including control of traffic from the west, preventing the railway boarding of infected persons, stationing doctors at key points on the railroad lines.

It was to no avail; the railway authorities simply closed down the line, dispersing the fleeing, panic-stricken people, who fanned out with their horses and donkeys and wagons and continued their flight.

Dr. Ekfelt, who had not been in China long, did not understand the people he was dealing with. He and Wu performed an autopsy at the home of a traveling salesman who had returned home from Kueihua two days before. Ekfelt, a pathologist, removed the spleen and left the body as it was, cut open, the clothes disarranged, soaked in blood. That night, a mob led by the dead man's enraged father tried to burn the wagon in which the doctors were sleeping, though fortunately all escaped with their lives.[63]

All in all, about sixteen thousand people died in the Shansi outbreak, which lasted until the spring of 1918.

The third, and far better-known, pneumonic plague outbreak came in Manchuria once again, in 1920. This time, the North Manchurian Plague Prevention Service was in place, and as Wu puts it, "This second visitation of the much-feared infection did not take the authorities by surprise, and they could deal with its various phases with vigor and confidence, relying on the knowledge gained during the previous eight years from their observations and research."[64]

Manchuria had remained plague-free for nearly ten years, though sporadic bubonic and pneumonic cases had continued to appear in Mongolia and Siberia. But in October of 1920, the wife of a Russian guard at Hailar in Manchuria died of the plague. So did three of her five sons, and then three Chinese soldiers billeted in the same compound. Wu and his team, oddly, were in Hailar at the time, and actually witnessed the illness of the guard's wife and the start of the epidemic. These initial cases were bubonic: Wu says that he and his associates witnessed the outbreak turn from bubonic and septicemic into pneumonic. Again, the principal means of spreading the infection were the "promiscuous spitting and huddling together" of soldiers and workers in their suffocating inns.[65]

Despite the local spread of infection, the medical oversight worked, to some extent, and only fifty-three died at Hailor. But the Chinese soldiers, rebellious now and restive, were far less willing than they had been ten years before to follow the orders of the plague control workers.[66] Several soldiers deliberately released a group of nine contacts, who fled. Two came to the crowded coal mining town of Dalainor, one hundred miles away. This really started the epidemic, says Wu; over one thousand out of four thousand Chinese miners died at Dalainor. (The Russian miners, housed in better-ventilated, more generous quarters, largely escaped.)

Other infected contacts came to Manchouli, where 1,141 people died of plague. Still other contacts came to Tsitsihar, where 1,734 died, and Harbin, which lost 3,125. Plague workers at Harbin, through the imposition of strict railway controls, managed to bottle up the epidemic somewhat, though some people still escaped and the infection continued to spread, eventually reaching the Russian city of Vladivostok by April 9. By the time that the outbreak ended in October, 520 people had died there.

The second Manchurian epidemic, all told, took the lives of 9,300, including six hundred Russians. Not only people in the crowded inns died, and not only in the winter. The outbreak lasted, says Wu, from the Russian guard's wife in Hailor to the last death in Vladivostok, exactly one year.

Using only strict quarantine and isolation, methods dating from the Renaissance, but wielded now with confidence and understanding, Wu and his workers stopped the epidemic and kept the death toll down. Wu insists in his various writings that there is no epidemic disease easier to control than plague, at least in theory. He could not treat it, and almost all plague patients died.[67] But he and his workers could control its spread.

* * *

One curious episode during the course of the plague outbreak at Harbin was to have a significant effect on the future of plague research. A ragged young Austrian Jew named Robert Pollitzer, who had graduated from the University of Vienna as a medical doctor, showed up at Wu's laboratory looking for work. He had been a prisoner of war, first of the Russians, later of the Japanese. He was utterly destitute and alone; and Wu, shorthanded, gladly put him in charge of the Harbin laboratory. This was an amazing stroke of fortune: Pollitzer was a trained pathologist who also spoke and wrote English, German, Russian, and French. He had tremendous initiative as well.

But Pollitzer suffered from intense loneliness and depression, and only the fortuitous visit of Wu and his associate, Chinese physician and researcher J. W. H. Chun, one evening stopped him from committing suicide.

> We then helped to find a wife for him from among a Polish refugee family, increased his salary and introduced him to more friends. The result was a remarkable change for the better in his disposition and habits. He became neat and punctilious in dress, was cheerful among others, and increased his capacity for serious work. Pollitzer certainly proved to be a most valuable asset to the medical staff and rendered invaluable service to us in Manchuria and later on in Shanghai on the establishment of the National Quarantine Service.

Pollitzer became one of the world's leading plague experts; his 1954 monograph on plague became one of the most influential texts on the disease ever written. For years he worked at the World Health Organization, until, in about 1959, he came to the United States. Robert Brubaker, the renowned plague guru, is perhaps the last plague expert to remember him personally:

> When I was young and stupid and just drafted into the army, which would have been right after the end of the Korean War, probably '59, I was sent to Fort Detrick and was put to work for Mike Surgalla in the plague laboratory there.[68] Surgalla was a very compassionate person, he was just an excellent boss, in the sense that he gave me a chance to develop skills and knowledge that I have relied upon and used right up to the present day. He also had a sense of history and a sense of the position of running the primary American laboratory concerned with plague. His lab, too, was concerned

with offensive biological warfare—this was the pre-Nixon era. We were working on ways to use plague as offensive weapon, though in reality we were just doing basic research. It was all publishable, nothing was that ominous.

When Pollitzer was brought or came to this country, Surgalla set up a program where Pollitzer would work all day in the Library of Congress, and once in a while we would come and sit at his knee and he would regale us with stories of his life and career. Young people just starting out are not typically interested in this sort of thing. They think a man like that is over-the-hill, that they can't learn anything from that old gaffer![69]

But it didn't take him very long to convince us otherwise.

Brubaker remembers being struck with Pollitzer's passion for the struggle against plague. He also has a faint but definite memory that Pollitzer and Wu had some sort of falling out. Apparently Pollitzer came to disagree with some of Wu's scientific views, especially Wu's "rather mystical"—as Brubaker puts it—belief that the plague germ undergoes some sort of "exaltation of virulence" during the course of an epidemic.[70]

Dr. Wu went on to a distinguished career in China until World War II, when he fled back to Malaysia. Though he was a "modern Chinese physician," as he put it, he kept to traditional practices in his own life, including polygyny. His first wife was a frail woman who had once had bound feet. Most of her children died in infancy. Wu set up a second household with another, stronger woman, who bore him a second family, overlapping with the first.[71]

Beginning in 1912, and for the next nineteen years, Wu served as the director of the North Manchurian Plague Prevention Service. Formed with the rather reluctant cooperation of Russia, Japan, and a number of other nations, it succeeded admirably, due in large part to the tireless effort and dedication of Dr. Wu. Dr. Frederick Eberson, physician, plague researcher, and colleague of Wu's, wrote that he "was the first and last great medical leader in China and one of the foremost in the field of public health under any flag."[72] The service, which eventually affiliated with the League of Nations Health Organization, became one of the best known public health organizations in the world. It boasted a number of first-rate physicians, including three Austrians, an Englishman, and two Americans. As Nathan puts it, "Through years of extraordinary turmoil in North Manchuria, the service had hung on, offering medical care on an apolitical basis, treating Bolsheviks, Kolchakists, Chinese Troops, and Hunghutzu raiders in quick succession."[73]

All that came to a brutal and shocking end after the Japanese invasion of Manchuria in 1931. Dr. Wu Lien-teh was taken prisoner, and released only under pressure from the British consulate. He fled back to China. The North Manchurian Plague Prevention Service disintegrated.[74]

* * *

The 1920–1921 Manchurian epidemic was the last major pneumonic plague outbreak, though smaller outbreaks continue in a part of the world still haunted by the most lethal plague of all.[75] At the same time, the rat-borne outbreak in India was dying down. Less fatal than the pneumonic plague of Manchuria, which killed almost everyone affected (Wu counted three survivors), it took far more lives. And *Orientalis* plague continues to be a threat: it moved to Vietnam, where it became especially deadly during the Vietnam War, killing thousands. It also entered Madagascar, which continues to suffer—in some recent years, several thousand people have died of plague. Sporadic cases flare in much of the rest of the world: in the United States, in Latin America, in Africa.

In India in 1994, the city of Surat witnessed a plague outbreak, and a mass hysteria, not seen since the early days of the Third Pandemic. Over 500,000 people fled the city.[76] What actually happened in Surat remains uncertain. Some sources claim that thousands of people contracted pneumonic plague. Inadequate medical response, improper storage of serum samples, the handing out of tetracycline like candy in the streets, all obscure what actually took place. All that is known is that the CDC researchers at Fort Collins confirmed the presence of plague in eleven serum samples. According to Gage, plague was probably present in Surat; several dogs and one Indian gerbil were confirmed serologically positive for plague infection. Other infections, including melioidosis, may have been present in the human population as well.

A later outbreak in India, in 2002, was more adeptly handled, with no massive number of suspected cases, and no mass panic.

* * *

It must be pointed out that *no laboratory tests* have ever confirmed, to my knowledge, a difference in virulence between marmot and rat strains. They all kill rats and guinea pigs, and in about the same time.[77] In the laboratory, all cats are grey: all plague strains kill, and kill effectively.

But in the greater laboratory that is the wider world, plague strains from different portals behave differently in human beings. Marmot strains have a much greater likelihood of going pneumonic. Marmots suffer from pneumonic plague themselves, and pass it to one another in their burrows. The three explosive human-to-human epidemics in the last century, as well as many other smaller pneumonic bursts in Manchuria, in Tibet, in China, that for lack of further exposure did not spread, were caused by marmot plague, as was the Black Death, in all probability.

Does marmot plague actually cause a different form of disease in people? Plague physicians and researchers who have seen pneumonic plague in the United States insist that people are not spitting blood until they are moribund. Only when there is blood in the sputum can pneumonic plague be contagious, and this doesn't happen in the United States until just before death. As the famous

Swiss-American plague researcher Karl F. Meyer put it: "Plague seems to be contracted in many cases when the infectee comes within close range of the coughing infector. Since the frequency of cough and the quantity of bacilli sprayed undoubtedly vary from case to case, and probably from time to time in the same case, depending on the nature and severity of the respiratory involvement, the infectiousness of pneumonic plague patients also probably varies greatly. Early in the infection, when the patient coughs little and plague bacilli in the expectorations are few or even absent, he is not likely to be highly dangerous."[78] But the six doomed workers on a k'ang in Chi-fu barely seemed ill, and yet were spitting blood. This suggests that people with marmot-derived primary pneumonic plague may remain mobile longer even as their lungs disintegrate, enabling them to travel, to work, to mingle with other people, and to spread the infection—before they die—a pattern that fits the Black Death as well. If a patient begins to transmit a disease while he is still mobile, that will dramatically increase the disease's likelihood of spread. Perhaps this is the key to marmot plagues strains' greater transmissibility: these strains could produce bloody, germ-filled sputum earlier in the disease's course. Or perhaps these strains are more effective in blocking the early onset of severe disease symptoms. In other words, perhaps strains that come into the human population from marmots are especially good at keeping down inflammation and other host immune responses while lung destruction continues unchecked. One or the other of these strategies probably makes pneumonic plague more and more transmissible as it passes from person to person: either bloody sputum develops while the patient is still ambulatory, or these plague strains are especially effective stealth agents.

When I traveled to Russia to see the old Soviet bioweapons institute at Obolensk, I was looking at ground zero for marmot plague research. The Saratov Anti-Plague Institute Mikrob, which used marmot strains for its research, provided Obolensk with stocks of plague germs, the most virulent they had.[79] Naturally occurring marmot plague is no longer an epidemic threat; antibiotics and quarantine successfully curtail short-term outbreaks. But the seed stocks from Saratov and Obolensk are still a danger, particularly when coupled with antibiotic resistance. The threat of a Fourth Pandemic remains.

# VII

# THE ENDURING THREAT

---

**This thing of darkness**
**I acknowledge mine.**

*THE TEMPEST*

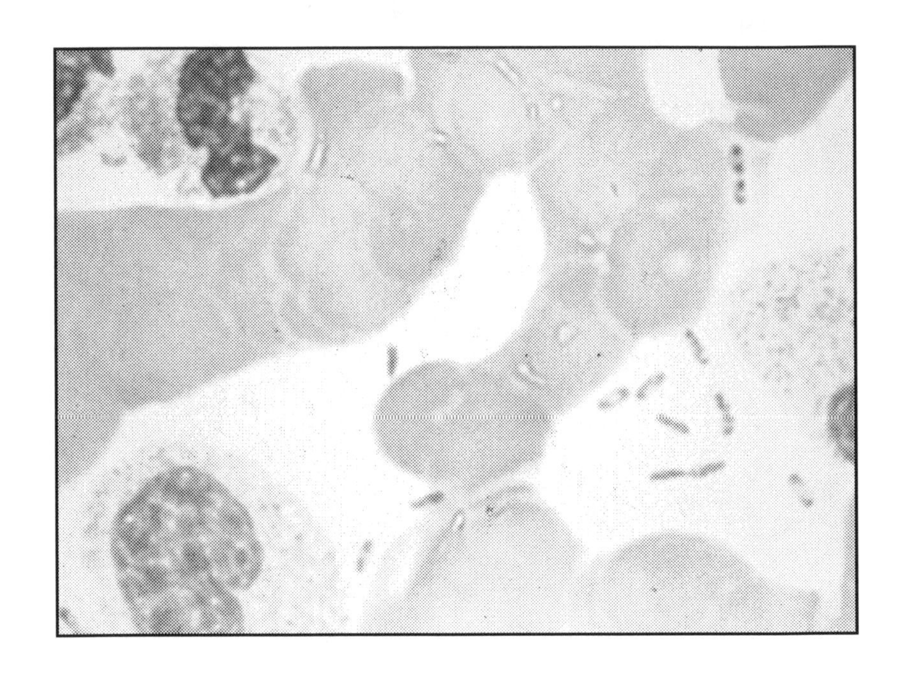

Blood smear showing lymphocytes and safety-pin shaped *Yersinia pestis* germs. COURTESY OF THE CENTERS FOR DISEASE CONTROL AND PREVENTION

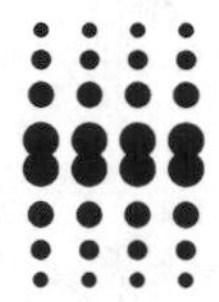

Plague fighter and plague warrior: the Janus faces of plague research. In the ambiguous figures of Igor Domaradskij and others, it is difficult to know where one aspect leaves off and the other begins. Even Robert Pollitzer, that grey, formal, dedicated man, who had spent so much of his life pursuing the plague demon from country to country, epidemic to epidemic, who worked for decades for the World Health Organization, even he came at last to the offensive program at Fort Detrick.

During the Cold War, Brubaker says, he had no doubt that plague research at the Fort Detrick bioweapons lab was intended to counter a mortal risk from a confirmed enemy. Pollitzer, having been a Russian prisoner of war, had no love of the Soviet Union, and willingly cooperated in helping to meet that threat. As Brubaker puts it, "Pollitzer knew Fort Detrick was a biological weapons laboratory; he was well aware this was offensive work. He had not been treated very well by Russia, and he understood that this country would not initiate BW. There wasn't any kind of moral issue—those were different times."[1]

Even so, there were always scientists on both sides of the Iron Curtain who did not participate in any kind of offensive biological weapons research. Vladimir Motin, former Russian plague researcher who now works at the University of Texas Medical Branch at Galveston, always worked in open programs, and, despite a personal respect and liking for Domaradskij, has come to view biological weapons research with horror. So does Michael Kosoy, a former Russian researcher now at the CDC, who refused to make the "deal with the devil," as he puts it, that many other Soviet scientists made. Still, both stress that nobody tried to make them join the System. On the American side, Peter B. Jahrling, the leading U.S. Army virologist, first arrived at Fort Detrick right after President Nixon had shut down the old U.S. bioweapons program. Many of the old guard were still there, and this made Jahrling, who openly despised bioweapons work, uncomfortable. He refused the opportunity even to learn from the old guard how germs were aerosolized, dried off, and made into weapons, though he acknowledges now that this might have been useful knowledge to have for biodefense work. But Jahrling wanted no contact with the Black Arts at all.

In any event, in the words of Pentagon analyst Andrew Weber, we should remember that the Soviet bioweapons researchers "never killed anybody."[2] There are degrees in bioweapons work, as in anything else: none of the American or Russian bioweapons scientists can be compared to the Japanese who worked on the infamous Unit 731 in Manchuria before and during World War II. There, under the regime of Lieutenant General Ishii Shiro, germ experiments rivaled Auschwitz in cruelty and horror.

After the last pneumonic epidemic in 1921, plague came back to north Manchuria, this time not as a natural outbreak, but as an agent of human design. The plague germ, among other agents, was used by the Japanese as a weapon of mass extermination. Thousands of Chinese, American, and Russian victims, whom the Japanese experimenters called "logs," were subjected to various bacteriological experiments. Some were injected with plague, anthrax, and other disease germs, and dissected while still alive and conscious. Some were prisoners of war; many were taken off the streets of Harbin. The once cosmopolitan, thriving town had been the home of White Russians, Jews, Koreans, Mongols, and Europeans from many countries, all wood now for Ishii's mill. Dragged off the streets and accused of various trumped-up charges, Ishii's logs were sometimes subjected to kangaroo trials and sometimes simply sent straightaway to Ping-fan's death factories.

Thousands of other people, mostly Chinese, were apparently killed in large plague epidemics started by various Japanese field trials, some of which involved dropping of porcelain bombs filled with plague-infected fleas. Epidemic wave after epidemic wave swept over the area; some plague outbreaks reached further into China. According to one scholar:

> in 1940, a series of epidemics struck Nongan county, 50 kilometers northwest of Changchuan [the site of Unit 100, another Japanese biological weapons experimental unit]. The origin of the epidemic is still uncertain. There is some evidence that the pestilence may have come to Nongan by accident. Several scholars believe that waste from the Changchuan facility somehow seeped into the underground water table, and spread as far north as Nongan. Others are convinced that rats escaped from Unit 100 laboratories, and brought plague with them to the infected region. Still others are certain that the Nongan county plague epidemic was nothing more than a BW field test undertaken by Unit 100.[3]

Plague cannot be spread in water; perhaps the disease was spread by escaped plague-infected rats, though the thought of lab rats escaping and marching across the countryside to a city fifty miles distant seems rather doubtful. We know the Japanese conducted plague field trials—so a deliberate bioweapons attack on the region seems the most reasonable explanation.

Another Japanese biological weapons base was established in the ancient Chinese city of Nanking in 1939, two years after the Japanese conquered that city. For two months beginning in December 1937, Japanese soldiers had run wild, slaughtering, looting, and raping: the infamous Rape of Nanking. Some twenty thousand women were raped, and 200,000 men cut down in the streets.

After the rioting subsided the Japanese still had not done with Nanking; they built their secret laboratory right in the heart of the city. They tested innumerable pathogens and toxins, especially plague, cholera, and typhus, which were cultivated in massive quantities. They also bred fleas: growing them by the hundreds in gasoline cans—to spread plague.[4]

In 1940 Nanking's plague-infected fleas, along with infected grains of wheat and cotton seeds, were dropped over the area of Ning-bo, causing devastating epidemics of plague and typhus. Japanese soldiers also dropped cholera vibrios, the rotating, comma-shaped bacteria that cause cholera, into lakes, ponds, and well water.[5] Domaradskij visited Ning-bo years later, and was shown a map by Chinese researchers indicating where the Japanese had dropped their flea plague bombs. "According to Chinese information, many people died in these attacks," he notes in his memoir.[6]

The actual number of people who died in Ishii's bioweapons attacks and experiments will never be known, but the toll of the dead may have reached six figures.[7] The Soviets made use of the knowledge garnered by Japanese scientists in their atrocious activities, but they weren't the only ones. America intervened in the process of postwar justice for the sake of the information on biological weapons the Japanese were willing to share, and helped the worst of the Japanese war criminals escape punishment.[8]

Bioweapons research advanced significantly in the half-century since Ishii's primitive but lethal recipes. Ishii failed to develop the technology to aerosolize plague; his attempts to build anthrax bombs fizzled. The Soviets were more adept at solving these early issues; by 1971, they had already produced an effective smallpox weapon, and tested it in the open air on Vozrozhdeniye (Rebirth) Island in the Aral Sea. As it happened, a research vessel was sailing on the Aral Sea about fifteen miles downwind of the test site. A young woman technician was the only person on deck; she inhaled some of the smallpox germs, and, despite an earlier vaccination, within two weeks had contracted the disease. She had already returned home to the city of Aralsk, where the disease spread to nine other people. Three of those victims, all unvaccinated children, died of hemorrhagic smallpox, the most virulent form. That this was a hot strain is clear; it overcame vaccine-induced immunity in six people, and produced fatal hemorrhagic smallpox in three. That the Soviets had learned to aerosolize the virus, harden it off, and deliver it over long distances is also clear.[9] All this was before the days of genetic engineering.

* * *

In 1973, two years after Aralsk, Victor Zhdanov and his deputy Igor Domaradskij began the drive for an entire restructuring of the Soviet bioweapons program. Lysenkoism, the doctrine of Marxist biology that rejected both Darwinian evolution and Mendelian genetics, had crippled Soviet biology, which lagged decades behind the West. Both Zhdanov and Domaradskij despised Lysenkoism, though Domaradskij had had to insert paragraphs of ideologically correct "rubbish" into his dissertation in order to earn his degree.[10] As designers of the bioweapons program, they were in a position to promote modern molecular biology and genetics in order to produce strains more suitable than natural ones for military purposes. Later, genetic engineering allowed Soviet scientists to augment plague and other weapons strains beyond the wildest inventions of the Japanese program.

In 1986, a year or so before Domaradskij left Obolensk, Sergei Popov—the gentle, faintly weary DNA synthesizer who now works with Ken Alibek in Manassas, Virginia, at a biodefense corporation—was brought in from Vector at Novosibirsk to do genetic engineering experiments on plague and other pathogens. As Popov puts it, "Plague was already a king in Obolensk. Hundreds of scientists studied plague in several of the institute's departments. The microbe was encoded as Agent No. 1 on the secret list of bacteria used to make biological weapons."[11]

According to Popov, though, Biopreparat believed that plague research in the institute was still inadequate and had to be boosted. Popov was placed in charge of Domaradskij's laboratory, from which the latter had been summarily removed. At Vector, Popov had worked under Lev Sandakhchiev, who Popov claims is responsible for, among other designs, a plan to create totally artificial viruses. Popov synthesized the DNA to produce peptides—small protein chains that form various human immune chemicals—and, with Vector virologists, inserted them into mousepox virus and vaccinia virus, stand-ins for smallpox. Popov's team, consisting of more than fifty experienced biochemists, had for eight years hand-synthesized various strips of DNA capable of regulating production of other proteins in artificial viruses. These were feats of almost unimaginable complexity; such experiments showed that, at least in theory, viruses could be genetically altered to express new, unexpected, and deadly properties.[12]

At Obolensk, Popov used the same methods to work on plague and other bacteria: he inserted synthesized DNA sequences into plasmids that could be delivered directly into the cytoplasm of the microbe's cells. These artificially augmented plasmids were designed to yield a variety of effects, including terrifying new forms of plague. In addition to plague, Obolensk scientists also created and tested variants of anthrax, glanders, legionella, and tularemia.

A patient infected with these genetically altered plague strains would first contract what appeared to be a typical pneumonic plague infection. The patient would immediately be treated with the appropriate antibiotics, which would cause the plague bacterial cells to break apart. The patient might appear to recover, but within his body, silently, Popov's new device would be working. The dying plague bacteria would release their cargo of artificial peptides, which in turn would cause paralysis, high blood pressure, irregular heartbeat, changes in behavior, perception of pain, or other bizarre effects. Two days, or two weeks, later, the recovered patient would be struck down by a heart attack, stroke, or instant and fatal paralysis. The experiments in animals confirmed that the approach worked as expected. "Recombinant plague," says Popov, "was as far away as an overnight incubation in the test tube."[13]

One scientist from Sandakhchiev's Vector Laboratories, Deputy Director Sergei Netesov, appeared one day in 1987 at Obolensk with a new idea for plague: he proposed taking the entire viral genome of Venezuelan equine encephalomyelitis (VEE), perforating the plague cell membrane, and planting the virus inside the plague cells like another plasmid. At Vector, Netesov had made a career of proposing such devices. Popov, quoting Alibek, identifies Netesov as the originator of the whole concept of chimeras: genetically engineered viruses made of two component parts—smallpox and VEE, smallpox and Ebola. Apparently on the strength of these novelties, Netesov had been promoted to deputy director of Vector at Novosibirsk. The name of the program he directed was *Okhotnik*—Hunter. His proposed plague-VEE chimera, fiendishly simple in design, but ferocious in concept, is probably the first time anyone had proposed putting together a bacterium and a virus. A victim of this chimera would be treated for plague with the appropriate antibiotics, which would kill the plague bacteria. But shattering the bacterial cell walls would release VEE directly into the lymph or the bloodstream; the invading virus would have already bypassed much of the immune system, and it would make straight for the brain. Within a week or ten days, the patient would be dead of encephalitis.

Domaradskij does not remember the meeting at which the Vector scientist proposed his idea. Either he had already been forced out of Obolensk by the time of Netesov's visit, or, deeply enmeshed in his struggle with Urakov, he did not participate in the meeting.

Netesov went back to Vector; according to Popov, Urakov later told him that "Okhotnik was successful." But Domaradskij is not impressed by Netesov's scheme: it would not work, he insists. The new VEE plasmid would quickly drop away—the trait could not be passed on for long among replicating bacteria. About Netesov, he says, "He's a molecular biologist—what do they know?"[14]

Domaradskij is a microbiologist of an older school. Unlike many biologists working today both in the United States and in Russia, he does not think of bacteria and viruses as strips of DNA, to be modified in any way you like. Some Western scientists propose that new techniques of "gene shuffling" can produce hundreds or thousands of new pathogens just by cutting and pasting DNA. But to Domaradskij, and to anyone trained in evolutionary biology, organisms are living things—you can't just chop and change one piece and expect the whole to continue merrily on its way. Viruses and bacteria, no less than butterflies, pigeons, apes, and molecular biologists, have been shaped by eons of natural selection, which, by definition, favors those strains best able to reproduce. Natural mutations are usually devastating for living things. Why wouldn't unnatural mutations be as well? If we think of a germ as an especially delicate watch, made by the most exquisitely precise of Swiss watchmakers, any change could throw the whole out of gear. A gene shuffler, no matter how sophisticated his machinery, would be like a mechanic trying to fix a watch with a claw hammer. Any genetic addition necessarily changes the germ in complicated and unpredictable ways. It was once thought that each gene has a single function, but we now know that one gene may affect an organism in several ways; it also may take two or more genes to produce a single discernible effect.

But Popov maintains that genetic engineering of bacteria or viruses is not really so difficult if you know what you are doing. "Unpredictability and complexity are a matter of knowledge. A gene shuffler could have the knowledge of the watchmaker and the intention of an engineer, wondering whether the watch could 'tick-tock' faster or slower. It is really easy, when you know which spring to change," he says.[15]

* * *

Still, other scientists remain skeptical that bacteria and viral genes can be combined in such a way. "I just don't buy this," says Vladimir Motin.[16]

I ask Domaradskij whether any of these genetically engineered strains would actually be transmissible person to person. For the plasmids for antibiotic resistance he himself had introduced, Domaradskij had devised a simple way of maintaining them as they cycled from person to person. The plasmids would necessarily drop away, he explains, unless you maintain constant pressure from antibiotics. Faced with a plague he did not know was antibiotic-resistant, a physician would reach for the one thing that would maintain the antibiotic resistant plasmids inside the plague bacillus: antibiotics. By treating his patient in the right and rational manner, he would be creating the very selection pressure necessary to ensure continued propagation of the antibiotic-resistant germ.

This device is simple, brilliant, and terrifying; we have no reason to think the strategy would not work. Given modern advances in genetic research, furthermore, scientists could probably dispense with plasmids altogether, as two Obolensk scientists, Andrei Pomerantsev ("the best of my collaborators at Obolensk," says Domaradskij) and Nickolai Staritsin, showed by integrating foreign bacterial DNA directly into the chromosome of anthrax. This might make antibiotic resistance even harder to dislodge, particularly with the pressure of antibiotics.

But Domaradskij does not have the last word on genetic engineering in plague; he was a pioneer, but he left the business in 1987. Popov remained at Obolensk until 1992, when the Soviet Union had fallen apart and the glittering, shoddy edifice of Obolensk began to crumble around him. Popov does not claim to be a plague expert, and though he inherited Domaradskij's laboratory, he did very different work. Much of his research, and some of what he claims are his most successful engineering experiments, involved the bacterium known as *Legionella*, the agent of Legionnaire's Disease. Genes added to produce paralysis worked best with *Legionella*, and not very well with the plague germ. But other people, including Urakov's old colleague and friend K. I. Volkovoi, found other genes that worked much better with *Yersinia pestis*.

According to Popov, Volkovoi's team successfully tried several approaches to introduce antibiotic resistance into plague, including the use of natural plague plasmids, integrating the resistance genes right into an existing plasmid.[17] But, despite the ostensible shutting down of the Russian bioweapons program by President Boris Yeltsin in 1992, Popov does not believe that Volkovoi's Obolensk plague research ended there; he claims that recombinant plague reports from Obolensk continue to appear.

Sometime in 1992 or 1993 Volkovoi and an associate reported the results of another genetic manipulation, an apparent attempt to create a still more virulent plague strain. Russian scientists define virulence as the number of microbes it takes to produce a lethal infection. Plague, one would think, is virulent enough: we know that under ordinary circumstances one or two germs may be enough to produce a lethal infection. But Soviet bioweaponeers consider that six hundred to fifteen hundred organisms are required per infection during an aerosol plague attack (anthrax, by contrast, was thought to require some 25,000 spores, though after the U.S. anthrax mail attacks it appears that far fewer spores may be necessary). These figures are all based on animal data. The figures come from "exchange volume": how many particles would be released into an aerosol chamber, how many would be expected, given the size of the chamber, to be inhaled. To figure this out, you have to factor in the strain, the type of preparation, the number of viable cells in a certain quantity of the preparation, and so forth. With the addition of diphtheria toxin, a deadly poison, Volkovoi and his coworker hoped to reduce the number of cells required to produce a fatal infection.

Instead of adding an additional plasmid containing the genes coding for diphtheria toxin, the two Russian researchers integrated the novel genetic information directly into one of the three natural plasmids. Integrating the code for diphtheria toxin directly into a natural plasmid, Popov says, guarantees that it will be stable. On a natural plasmid, furthermore, it is much more suitable for expression and manipulation than it would be on the plague germ chromosome itself.

They hoped for greater virulence; what they got was surprising, and far more ominous. When they tested their plague-diphtheria chimera on monkeys, they found that it overcame immunity to live plague vaccine. In other words, the new plague chimera was, at least to some extent, not only lethal at lower doses than normal weaponized plague, but also vaccine-resistant.

A hyper-lethal, vaccine-resistant, antibiotic-resistant plague weapon would be the most powerful biological weapon of all—a veritable Andromeda Strain, a Black Death for the twenty-first century. There is no reason to think such a weapon has ever been made. But the technology may now exist to produce it.

* * *

In the United States, the main biological weapons threats are considered to be smallpox and anthrax. But to the Russians, plague is the most dangerous bacterial threat agent.[18] As anthrax and plague genomics expert Paul Keim of Northern Arizona University puts it, the Russian and U.S. biodefense programs are like mirror images of each other: we fear anthrax, since it is both lethal and extremely durable in the environment, and they fear plague, for its virulence and transmissibility.[19]

Part of the reason for this discrepancy is that when American biodefense experts think about plague, they view it in the light of American experience, where plague has never proved very contagious. Thomas Inglesby of the Johns Hopkins Center for Civilian Biodefense, in an article summing up the perceived risks of anthrax and plague as a biological weapon, puts it this way:

> Outbreaks of pneumonic plague have historically been uncommon. The likelihood of person-to-person contagion appears to be dependent on both patient and environmental factors, although patient factors are less well understood. The environmental factors supporting spread would include cold temperatures and close, cramped, crowded conditions. . . . The largest pneumonic plague outbreak was in Manchuria in 1910 and 1911, in which there were 60,000 cases of pneumonic plague. . . . The last case of person-to-person transmission of pneumonic plague was in Los Angeles in 1924, when 1 person with secondary pneumonic plague gave primary pneumonic plague to 32 close contacts. . . . None of the pneumonic plague cases occurring in the United States since that time has resulted in a single case of transmission of plague.[20]

It is the familiar mantra: the only really large pneumonic plague epidemic in history occurred in Manchuria during 1910–1911, when, as Brubaker says, "it was cold and they were all breathing in each other's faces." Never mind that the evidence for the role of pneumonic plague in the Black Death is overwhelming; never mind the sixteen thousand dead of Shanshi, China, or the ten thousand who died in the second Manchurian epidemic. Never mind the clear evidence of explosive contagion when marmot plague breaks into human communities.

Russian and other former Soviet plague experts, for their part, know suslik plague, gerbil plague, and marmot plague; the last is the most lethal and the most transmissible, but all of these natural foci are old, and all of them are dangerous. Drs. Atshabar and Suleimenov of Kazakhstan warn that certain American oil companies with interests in Kazakhstan are careless in their penetration of some of these natural plague foci; they fear an epidemic among heedless American workers. The Kazakh scientists themselves watch plague outbreaks very closely; whenever a single case crops up, dozens, sometimes hundreds, of people are hospitalized and watched, which strikes American observers as strange indeed. But the Kazakh plague experts understand the threat very well; their plague foci should not be confused with the prairie dog towns of the American West. The fact that athletes and children who run through the prairie dog towns in the United States never seem to come down with plague should not blind us to the risk of lethal, transmissible plague to oil workers in rural Kazakhstan. This is a matter of real concern to Suleimenov and Atshabar; they fear a major outbreak, which could be brought into the United States by oil workers who return home before they fall ill.

Another reason that Russian biodefense experts—and former bioweaponeers—fear plague is that they know what has been done with it. Even apart from genetic engineering, the process of turning plague into a weapon involves hardening off the germs until they can survive, at least for a while, in the external environment. Plague germs are not nearly as resilient as anthrax, which forms a spore almost impervious to heat, cold, desiccation—to anything but ultraviolet light. Although plague germs can survive in the external environment for a time when cloaked in sputum, or frozen in a corpse, or buried in rodent burrows, in the atmosphere it is thought that no weaponized plague strain can last longer than about an hour or two. Popov says that would be long enough. Under the right conditions, delivered in the early morning over a city, or sprayed from an atomizer into a stadium or subway, plague germs could cause a vast and lethal outbreak, made worse since the disease could spread.

In a sense, plague makes a bioweapon all the more hazardous for being short-lived. An invading army would not have to worry about the decontamination of surfaces, as they would with anthrax spores. The decontamination of buildings during the 2001 anthrax mail attacks in Florida, New York, and Washington—which involved minuscule quantities of spores—show what a nightmare that process can be.[21] A lethal agent that conveniently disappeared a few hours after dispersal is much less hazardous than anthrax to soldiers entering an area after an aerosol attack.

Plague was never central to the former U.S. bioweapons effort, for one simple reason: despite sophisticated research on plague aerosol infectivity, among other topics, carried out at the old bioweapons laboratories at Fort Detrick before the 1969 shutdown, American scientists were never able to grow the germ in bulk. "We could grow it in a high concentration for laboratory work," says William Patrick III, a leading bioweapons scientist in the old U.S. program. "But in large-scale quantities, it lost its ability to infect. But we did not want to use contagious agents anyway, because of the risk that they'd go where you didn't want them to. So we didn't try real hard to overcome the problem."[22]

There is a trick to growing virulent plague in bulk, which the Soviets (and even, apparently, Ishii's men) well understood. But our scientists did not know the trick. When Domaradskij heard of this, he seemed astounded. Then he laughed and said, "Let them come here, we'll teach them!"

Despite this failure, the old work at the British and American bioweapons programs has added considerably to our store of knowledge about the workings of plague infection. None of this work—for good reason—could be done today. But the aerosolizing experiments done on monkeys at Porton Down, the British bioweapons research site parallel to the American Fort Detrick, in the 1940s and 1950s may help us, in a roundabout way, to understand one of the continuing mysteries of natural plague outbreaks: why they come to an end.[23]

In a series of articles, Wu Lien-teh and some of his fellow researchers proposed that pneumonic plague epidemics end because of what was then called "an exaltation of virulence." As plague germs, during a pneumonic epidemic, continued to be passed from lung to lung, they were thought to grow ever more virulent, until at last they killed too quickly to produce a real pneumonia, but produced instead what Wu called "pulmonary plague." Pulmonary plague, showing fatal lung changes but no pneumonia, killed before lung tissue destruction could be advanced enough to produce cough and blood-spitting. Without that, pulmonary plague could not spread to other patients. Toward the end of several outbreaks, Wu and his colleagues noted that the number of so-called pulmonary cases, fatal within hours, seemed to increase; he suggested that this natural process of rising virulence brought plague epidemics to an end.

From an evolutionary perspective, his theory seems unlikely. There ought to be strong natural selection on plague germs to keep the disease within the bounds of virulence. K. F. Meyer, a leading plague scientist in California during the 1950s and 1960s, claimed that cases of so-called pulmonary plague were not principally lung plague at all, but actually a form of bubonic plague with *internal* buboes. These formed near the upper respiratory tract. Apparently these patients breathed in plague germs that somehow landed on the tonsils or other mucosal membranes in the throat, where they began to grow, producing regional buboes that could not be seen, and rapidly developing into septicemic plague.[24] But why? Why didn't these inhaled germs settle in the lungs?

The old work of Porton Down's bioweaponeers and similar research solves this puzzle.[25] It turns out that the size of the particles inhaled determines the form of the disease in a particular patient.

To produce pneumonic plague, the inhaled plague particles (which, in a bioweapons experiment, would consist of plague germs hardened off and coated in some fashion) *must* fall within 1.5 to 10 microns in order to penetrate deep into the lungs. If particles are too small, less than one micron, they fly into the lungs and out again, like gas or air. If they are larger than 10 microns, the plague germs settle in the *upper* respiratory tract, where they infect the tonsils or the lymphatic system and produce rapid septicemic infection. Septicemic cases in a small population may indeed help to bring the outbreak to an end, as fewer patients will be contagious, but this is merely a fortunate accident. As Meyer puts it, "An explanation of this unusual [pulmonary] form and the better-known pneumonic form began with the work of Wells and Wells and many others on aerial infection with droplet and droplet nuclei. The wide differences in behavior between large and small droplets is obviously significant in these forms of plague. . . . On first thought, it might seem that such particles would enter the respiratory tract and be deposited at resistant sites as in other bacterial pneumonias. But the upper part of the respiratory tract is furnished with no known defense against *P. pestis.* The smaller the particles are the more likely they are to be inhaled and lodge in the highly susceptible lung itself. . . . Perhaps only the larger particles can lodge in the upper respiratory tract and give rise to tonsillar or septicemic plague."[26]

Particle size is something the Soviet researchers well understood. The plague warriors designed their weapon with those size requirements in mind. They knew that to produce a transmissible lung infection, particles between 1.5 and 10 microns were the ideal size.[27]

Former Soviet plague warriors have held, for a long time, the keys to several mysteries of plague. They understand plague differently than our scientists do, because their experience with plague is different. They understand that it is the reservoir host that determines how virulent a plague strain will be, for the obvious reason that in certain ancient strains the arms race between host and germ has gone on for longer, and that the disease has of necessity armed itself with greater virulence as the host population's resistance increases. They understand that the most virulent plague of all is found in Central Asia, among the various species of marmot. They know that marmot plague moves directly to the lungs, and that therefore it is the best form to produce pneumonic disease in men. They know that the form of disease produced is determined by the size of the particles used in an aerosol suspension. They also know what can be added to plague to enhance its virulence, or give it antibiotic resistance.[28]

Putting all of this together gives us a frightening picture quite unlike that of plague as it is understood in America. The potential for a lethal weapon, another Black Death, is still there, in the stocks and storehouses of the former Soviet Union. Supposedly the actual weaponized stocks were destroyed, though no one in the West has seen evidence of that destruction. The Soviet bioweapons program was ordered shut down by President Yeltsin in 1992, and no one expects that weaponized stocks of plague from that era could still remain viable. But the seed strains remain, and the knowledge remains. Yeltsin could order an end to plague and anthrax and smallpox production, but he could not erase the requisite knowledge from his scientists' minds and hearts.

Obolensk is crumbling; as I write this in August 2003, General Nikolai Urakov, Domaradskij's former nemesis, has recently been removed from his post because of his mismanagement of the former All-Union Institute of Applied Microbiology. There are efforts now underway to remove all dangerous pathogens from the institute and to convert the entire complex, with the help of U.S. funds, to the production of drugs and medicines. But four other former bioweapons laboratories, which continue to be operated by the Russian Ministry of Defense, are still completely closed to the West. They are black holes; we have no idea, and Russians like Domaradskij have no idea, what goes on in those laboratories. If bioweapons work does not continue, why does the Russian Ministry of Defense continue to keep American delegations out, and to prohibit contact between Ministry of Defense and American scientists? Many American scientists are still concerned about what goes on at Kirov, at Pokrov, at Sergiyev Posad, where smallpox virus was once grown by the ton, and at Ekaterinburg (formerly Sverdlovsk), the site of a 1979 bioweapons accident that left sixty-eight people dead of inhalational anthrax.[29]

The strains exist, the knowledge exists. We are not in any apparent danger of a bioweapons attack from Russia itself. But many former Soviet scientists are unemployed, and some of them may be hungry. Quite a few of the best specialists have made their way to the West; but there are others.

Popov speaks with considerable bitterness about the world he left behind in 1992, when he fled to Britain: "It was so miserable to be a scientist in Russia; no money, no status." Urakov's sharp dealings kept his staff on the edge of starvation; he forced them to buy sugar from the enterprise he controlled at exorbitant prices. The scientists were sometimes reduced to desperate measures just to eat.

One winter night wolves from the deep woods that surround Obolensk broke into the hutches where the scientists kept their experimental rabbits, which were not infected, but which had been immunized against a whole range of bacterial diseases. The wolves tore out the rabbits' throats and left them. In the morning Popov woke to see the dead rabbits lying in frozen pools of blood on the snow. He thought to himself, "The meat is undamaged, and there's enough food for six months." He issued a permit for the carcasses to be put in garbage bags and transported outside the facility. But at night in his garage, he skinned them and put them in his freezer. Those rabbits fed his family a whole winter long, while Popov went to England to look for work.

Is it any wonder that some Russian scientists have peddled their knowledge to the highest bidder? Some have gone to Iran to share their expertise; others to other countries, including Iraq. One scientist, N. Kislichkin, opened his own company, Bioeffect, at Obolensk, with offices in Vienna and Moscow. He tried to hawk Domaradskij's own genetically enhanced, antibiotic-resistant tularemia strain, though it is unclear whether there were any takers. He produced a commercial flyer, which, in Domaradskij's translation, offers "to create novel microorganisms of a vaccine group for infections on the basis of a customer's order. Bioeffect was ready to cooperate in research activities within investigations of virulence factor of different infections."[30] Kislichkin characterizes his business as follows:

> Yes, I really have my own private biological firm. But I am working absolutely legal and creating genetic changed microorganisms only vaccine groups (and already obtained and checked new vaccine candidates). Besides, I have own Russian patent 1 2085584 "Methods of obtaining of recombinant tularemia strains, which produced of virulence factor" for my work.[31]

The fact that Kislichkin may only be selling vaccine strains does not mean very much: the technology to create genetically altered vaccine strains and pathogenic strains is exactly the same. Vaccine strains in the Soviet bioweapons program were often used as stand-ins for dangerous strains.

This is the fear that keeps some U.S. experts up at night. There is no accounting for the movement of strains, or of vials; there is no accounting for the movement of scientists. The Soviet bioweapons system, at its height, employed tens of thousands of workers, some thirty thousand in Biopreparat, the so-called civilian program, alone. Of those workers, several thousand were genuine research scientists. Where are all of them now? No one knows.

* * *

Natural plague, as we have seen, does not pose much of a threat to most of the world today, except perhaps in the oil fields of Kazakhstan. But even in those areas that suffer most from plague outbreaks, such as Vietnam or Madagascar, rampant pneumonic epidemics do not occur. Those cases can be handled; the most recent plague outbreak in India, in 2002, was quickly controlled. Two bubonic plague cases in New York were also swiftly recognized and treated; these patients, a husband and wife, came from New Mexico, "the land of the flea and the home of the plague," where rodent plague is common. The likelihood of spread from those two cases was nil.

Rat plague presents few problems to the world as a whole, devastating though local outbreaks may still sometimes be. Marmot plague is a greater threat, but, carefully watched, and with no chances for long chains of human transmission, it, too, does not represent a Black Death waiting to happen. Plague—from any portal—cannot enter the human species without intimate human participation: there must be ways to transmit the disease readily, through the transportation of rats and their fleas, or through human fleas jumping from person to person, or through lung-to-lung transmission: infected people must be able to travel easily for pneumonic disease to spread. Public health officials can contain it just as they contained SARS, by rapid response, quarantine, and careful monitoring, and the use of face masks.

Today, plague is one of the three reportable infections: countries with plague outbreaks are required to notify the World Health Organization. Even the smallest outbreaks, involving one or two people, are handled in Kazakhstan, for example, with what seems to Americans overkill: scores of people were taken to the hospital for observation when plague broke out in a community. It is not overkill; the Kazakh scientists know exactly what sort of plague strains they face, and they take no chances. Even if, by some chance, plague infected the rats of New York City, insecticide and antibiotics could swiftly handle the situation.

But a deliberate plague attack is something else again. Ishii's porcelain bombs with their cargo of fleas and germs killed thousands—what could a sophisticated, aerosolized plague strain do, released over a major city? An influential article in the *Journal of the American Medical Association* published in 2000 cited a 1970 World Health Organization analysis that suggested that "in the worst-case scenario"

> if 50 kg of *Y pestis* were released as an aerosol over a city of five million, pneumonic plague could occur in as many as 150,000 persons, 36,000 of which would be expected to die. The plague bacilli would remain viable as an aerosol for 1 hour for a distance of up to 10 km. Significant numbers of city inhabitants might attempt to flee, further spreading the disease.[32]

The WHO's analysis was done in the happy days before genetic engineering, before antibiotic-resistant plague. Our present worst-case scenario is much grimmer. What could we do in the face of a plague attack with marmot plague strains, with built-in antibiotic resistance, carefully designed for maximal penetration into the deep recesses of the lungs? Antibiotic-resistant plague strains have emerged in nature; twice, in Madagascar, thanks to antibiotic-resistant plasmids obtained through lateral transfer from other bacteria (probably from *E. coli*). Fortunately, they did not spread, and the new plasmids never became established.[33] The threat from genetically engineered plague strains is much greater; any such strains would be delivered in aerosol form. As we have seen, Russian scientists always left in some antibiotic vulnerability so that they could treat their own scientists in case of a laboratory accident. Chances are that treatment could be found in time. But in how much time?

Alibek insists that the old Soviet program at Obolensk, using Domaradskij's method, was able to produce weapon strains resistant to about ten common antibiotics.

In the case of an outbreak, we could not even be vaccinated against the spread of the disease, as there are no killed strains or engineered vaccines that work against pneumonic plague. There are candidates being tested, both at Porton Down in the British biodefense laboratories, and at USAMRIID, the United States Army Medical Research Institute of Infectious Diseases, but no vaccines are currently available. The United States tends to avoid the use of live vaccines; in American eyes, the live EV strain used in Russia is unacceptably dangerous, often producing considerable swelling and fever. Worst of all would be some type of vaccine-resistant plague. All in all, we cannot have much confidence in the use of vaccines to stop an epidemic if a bioweapons attack took place.

In the absence of some novel treatment, a new antibiotic, a new vaccine, we would be plunged backward in time, not by any means to the days of Gui de Chauliac, but to the era of Wu Lien-teh. Terrifying though such a scenario would be, it could be handled again the way Wu Lien-teh and his fellows once handled the massive Manchurian plague. Wu had the benefit of a docile population during the first epidemic, though by the second the local Chinese, particularly the soldiers, had grown restive and difficult to handle. But a docile population is not what would be needed in a crisis of this magnitude. People would need to understand how to protect themselves; the government would need to be able to communicate simply and believably to its citizens.

In the time of a new plague, people would need to remember, despite their terror, that *no epidemic disease is more susceptible to quarantine.* The very notion of quarantine has an ancient horror; in the brutal lockups of the Renaissance, people starved to death, or died of disease in miserable isolation. But today, people would need to impose upon themselves the requirement, in the event of an outbreak, to *stay out of public places, and, if at all possible, to stay in their homes*. We would need gauze masks to run errands; we would need access to food, water, and medical care. None of this would be easy to arrange. But terror and flight have always been plague's handmaids. We would need to resist the age-old desire to flee, to head for safer country somewhere else, in the hills, in the woods, in some other city. We would need to remember that, unlike anthrax, plague cannot linger in the environment. A plague attack would be over long before the first infections ever appeared. Therefore, flight makes no sense. Self-imposed isolation does.

In the advent of sickness, people would need either to be admitted to a hospital ward, or, if there are no beds, to be cared for at home, and the people caring for them would have to wear masks.

The power of epidemic plague can only be broken if the chain of transmission is severed. That is why, to successfully resist a genetically altered, antibiotic-resistant infection, to preserve the social order from being shattered as it was in Justinian's day and during the Black Death, both the government and its citizens would have to coolly implement difficult and painful restrictions on their own behavior.

May such a terrible attack never come to pass. But if it does, to combat a twenty-first-century plague, we will need to be twenty-first-century citizens.

# NOTES

## INTRODUCTION

1. Ewald, Paul W. 1996. *Evolution of Infectious Disease.* Oxford: Oxford University Press; and Ewald, Paul. 2000. *Plague Time.* New York: Free Press.

## I. RETURN TO OBOLENSK

1. The book, now called *Biowarrior*, was published in August 2003 by Prometheus Books (Amherst, New York).

2. If there were still volleyball courts, I did not see them.

3. There was also a parallel system of military laboratories, run directly by the Ministry of Defense; Domaradskij was connected closely to the military laboratory at Kirov for a number of years. To this date (September 2003) no Westerner has ever been allowed access to these four military laboratories: Kirov, Sergiyev Posad, Pokrov, and Compound 19 at Sverdlovsk/Ekaterinburg. And no one knows what research is still carried out there.

4. According to Domaradskij, Zhdanov actually came from Kharkov; he did not move to Moscow until age forty.

5. One hundred forty-seven nations have signed the convention to date.

6. Domaradskij, protected for years by a powerful patron, General V. D. Belyaev, was able to maintain this laboratory as an independent fiefdom for years, but eventually lost control of it some time after the 1979 death of Belyaev.

7. At his Moscow laboratory, Domaradskij worked with EV, an avirulent (nonlethal) strain of plague. But he proved the principle—that plague strains could be made antibiotic-resistant. The eventual weapon strain was made with Domaradskij's techniques by other scientists.

8. According to Alibek, though, and unbeknownst to Domaradskij, Urakov later adopted the despised binary method to produce weaponized plague. This plague, says Alibek, was resistant to about ten antibiotics.

9. In other words, giraffes' necks grew longer because they stretched to reach higher leaves; these longer necks were passed directly on to their offspring. Trofim Lysenko was a peasant farmer who gained astounding and terrifying control over all biological science in the Soviet Union; a protégé of Stalin's, he kept Soviet biology behind the West for years. Those who dared to disagree could be fired, or even executed, for their heresy.

10. Alibek describes one of these altercations between Domaradskij and Urakov, which involved shouting and streams of insults, in his book *Biohazard.* 1999. New York: Random House, p. 162.

## II. THE MYSTERY OF PLAGUE

1. Domaradskij, deep in his secret world, had made the same discovery three years earlier—but he could not publish it. Domaradskij and Orent, p. 182.

2. Domaradskij, I. V. *Chuma: Sostoyanie gipotesie, perspectivie (Plague: Situation, Hypotheses, Prospects).* Saratov: Saratov Medical Institute Publishing, 1998.

3. Septicemic plague is a severe form of the disease, with a high number of bacteria in the blood, producing a kind of blood poisoning, or septicemia.

4. Dr. Douglas Gray, who witnessed the great pneumonic plague outbreak in Manchuria in 1910–1911, writes: "There were no marked prodromal symptoms. Often a man had normal pulse and temperature and the next day was dead." Gray, G. Douglas. 1911. "A Report on the Septicemic and Pneumonic Plague Outbreak in Manchuria and North China" (Autumn, 1910–Spring 1911). *Lancet* 1(2):1152. These descriptions of sudden death are remarkably common in accounts of the Justinian Plague and the Black Death as well.

5. Ewald, *Evolution of Infectious Disease*, pp. 35–55.

6. Other fleas show blockage as well, but *Xenopsylla cheopis* is considered the most efficient vector.

7. Since a molecular study showing the close similarity between *Y.*

*pseudotuberculosis and Y. pestis*, and arguing that *pestis* branched off from *pseudotuberculosis* a relatively short time ago (Achtman et al. 1999), the belief that plague is a new disease has become widespread among specialists—but Domaradskij insists that this is only one, as yet unproven, hypothesis, and should not be accepted as a given.

8. Hinnebusch explains that the genetic signature for murine toxin genes is clearly different for these toxin genes than for the rest of *Yersinia* DNA. This tells us that at some point in the last twenty thousand years some other bacteria donated the genes for murine toxin to the plague germ. "These exchanges are very common in the microbial world," says Hinnebusch. "Bacteria tend to pass DNA back and forth. They have two or three ways of doing this. They can use plasmids—this is the discovery Joshua Lederberg made for which he won the Nobel Prize" in 1951. Plasmids are strips or rings of extra chromosomal DNA found in the cytoplasm of many bacteria. "When two bacteria come into physical contact," says Hinnebusch, "the donor can inject DNA to the recipient. This is plasmid conjugation. Then there are the viruses known as bacteriophages, which can carry along with them some of bacterial DNA from the host cell they came from. The third method is called transformation—the bacteria picks up naked DNA from the environment—from water or soil."

9. According to Hinnebusch, this analysis comes from Robert Brubaker.

10. As Perry, Tom Schwan, and Hinnebusch discovered in experiments that took the gene for HMS out.

11. According to Domaradskij, HMS genes are also found in pseudotuberculosis germs, where they perform different functions.

12. Gage, personal communication with the author; Hirst, L. Fabian. 1953. *The Conquest of Plague: A Study in the Evolution of Epidemiology*. Oxford: Clarenden Press, pp. 238–41.

13. As Gage points out, not all of *Pulex*'s hosts really have nests.

14. The great plague researcher Robert Pollitzer puts it this way: "No doubt can exist that a transmission of plague may be effected through mass attacks of fleas with contaminated proboscides." Pollitzer, R., 1954, *Plague*. Geneva: World Health Organization, p. 350.

15. Williams, Peter, and David Wallace. 1989. *Unit 731: The Japanese Army's Secret of Secrets*. London: Hodder & Stoughton, p. 250.

16. There are certain kinds of fleas that can ingest plague germs and not

form blockages—the germs can live and grow in the flea's intestines. These unblocked fleas are actually not well investigated, according to Gage, and may still form a third mode of flea-borne transmission under certain circumstances. When I use the term unblocked fleas, I am not referring to these fleas; instead, I am referring to mechanical transmission via contaminated mouth parts of fleas such as *Pulex irritans.*

17. An institute that Igor Domaradskij headed for nine years, beginning in 1959.

18. Sergei Balakhonov, personal communication with the author. Other scientists, including Vladimir Motin, do not agree with this point.

19. That notion is quite old: it dates from a paper in 1972, and was accepted as dogma, Motin points out, for years. Recent molecular evidence, however, suggests that this old dogma—and the evolutionist intuition that such a widespread trait cannot be trivial—is probably correct. A March 2002 paper by a team of Swedish researchers suggests that, along with certain other genetic factors, the plague capsule serves to reduce the immune system's ability to phagocytose *pestis.* It appears to do this by interfering somehow with the communication between immune cell and bacterium. Du, Y., Rosqvist, R. Forsberg, A. 2002. "Role of Fraction 1 Antigen of *Yersinia pestis* in Inhibition of Phagocytosis." *Infection and Immunity* 70(3):1453–60.

20. As Tom Schwan of Rocky Mountain Laboratories points out, personal communication with the author.

21. Kartman et al. 1958. "New Knowledge on the Ecology of Sylvatic Plague." *Annals of the New York Academy of Sciences* 70(3):668-711.

22. Yops, it turns out, aren't merely the proteins found on the outer envelope of *Yersinia*, but the odd handle has stuck.

23. *Pestis* has three plasmids, one of which it shares with the other *Yersinia*s. Adding additional plasmids is how, initially, Domaradskij and his co-workers were able to add antibiotic resistance factors to plague.

24. Some scientists who specialize in these Yops spend a great deal of time worrying over whether the poison Yops drift down their needles in the form of proteins that are folded or unfolded—arcana that need not detain us here.

25. Vladimir Motin, personal communication with the author.

26. Interview with Elisabeth Carniel, September 2002, Turku, Finland. Also, Baltazard, M. "La conservatien de la peste en foyer invertere," *Médécin et Hygiène*. 1964, vol. 22. Pp. 172–74, quoted in Suntsov, V.V., and N.I. Suntsova. 2000. "Ecological Aspects of Evolution of the Plague Microbe *Yersinia pestis* and the Genesis of Natural Foci." Biology Bulletin 27: 6, pp. 541–52. Translated from *Izvestia Akademiei Nacek, Seriya Biologicheskaya*. 2000. No. 6.

27. Interview with Sergei Balakhonov, August 2003, Flagstaff, Arizona. But Suleimenov and Atshabar deny that plague survives in the soil in Kazakhstan; perhaps conditions are too hot and too dry.

28. Garnham, P.C.C. "Distribution of Wild-Rodent Plague." *Bulletin of the World Health Organization* 2:271–78, 1949; quoted in Butler, Thomas. 1991. *Plague and Other Yersinia Infections*. New York and London: Plenum Medical, p. 48.

29. Personal communication with the author. John Hoogland and I go back a long way. In 1972, Hoogland was a teaching assistant in a class I took in evolutionary ecology, taught by Richard D. Alexander of the University of Michigan Museum of Zoology.

30. Personal communication with the author.

31. As Domaradskij points out, however, there are other points of view: some scientists have argued that plague may have come to America a long time ago via the Bering Strait. But the rapid die-off of prairie dogs, the low level of plague virulence vis-à-vis other species, and their extreme sensitivity to plague seem to suggest a late introduction.

32. Dickie, Walter M. 1926. "Plague in California, 1900–1925," in *Proceedings of the Forty-first Annual Meeting of the Conference of State and Provincial Health Authorities of North America*. American Health Congress Series, Vol. 3.

33. December 1918 bulletin of the State Commission of Horticulture, Rodent Control Division, quoted in ibid.

34. Link, Vernon B. 1955. *A History of Plague in the United States*. Public Health Monographs, Public Health Service Publication No. 392. Washington, D.C.: United States Government Printing Office.

35. Personal communication with the author.

36. Strong, 1912. "Studies in Pneumonic Plague and Plague Immunization." *Philippine Journal of Science* 7.

37. Wu Lien-teh and Eberson, F. 1917. "Transmission of Pulmonary and Septicaemic Plague Among Marmots." *Journal of Hygiene* 16:1.

38. Though there is a curious article, in Russian, found by Igor Domaradskij, written by an early Russian plague researcher, who claims that when rats are experimentally induced to inhale plague germs, the strain quickly develops a tropism for lung tissue.

39. This classificatory scheme classifies plague strains according to whether the strains ferment, or acidify, a chemical called glycerol, or whether they do not (glycerol-negative strains), and also whether they can or cannot reduce nitrates to nitrites. Some researchers have proposed that glycerol-positive strains may have something to do with hibernation, while others, including Vladimir Motin, are skeptical; for the nitrate reduction trait, no one to my knowledge has yet proposed an explanation.

40. Recently, Motin et al. have argued from analysis of the genetic fingerprinting of plague strains that particular strains from the Caucasus Mountains are phylogenetically the most ancient of all *pestis* strains known. These strains entirely lack the pPCP plasmid, on which is encoded pesticin and the plasminogen activator, or PLA. They also cite two Russian scholars, Bobrov and Filippov, who on the basis of a different analysis came to the same conclusion. Therefore it might be most correct to say that the oldest *true* plague focus on earth is found in Central Asia. These intermediate strains, sometimes called *Pestoides*, actually tend to be less virulent than other plague strains. They circulate among voles. Vladimir L. Motin et al. 2002. "Genetic Variability of *Yersinia pestis* Isolates as Predicted by PCR-Based IS100 Genotyping and Analysis of Structural Genes Encoding Glycerol-3-Phosphate Dehydrogenase *(glpD)*." *Journal of Bacteriology* 184(4):1019–27.

41. Personal communication with the author.

42. Strains from the great gerbil of Central Asia are also thought by Suleimenov and Atshabar to be highly dangerous; these rodents are implicated in the Turkmenistan outbreak of 1950 described in Chapter 1.

43. Vladimir Motin, personal communication with the author. Martinevsky worked at the same institute in Kazakhstan as Suleimenov and Atshabar, but he has recently retired. Martinevsky's claim suggests that the vole strains of the Caucasus, known as *Pestoides* to indicate their distinction from true *pestis*, may not be virulent for people.

44. Suleimenov and Atshabar claim that plague among great gerbils also can cause explosive outbreaks among human beings—witness the Turkmenistan epidemic described in Chapter 1.

## III. THE WINEPRESS OF GOD

1. *Tanakh, The Holy Scriptures: The New Jewish Publication Society Translation According to the Traditional Hebrew Text*, p. 424.

2. Hemorrhoids are not fatal, nor are they the symptom of any infectious disease. The usual diagnosis of the Ashdod plague is bacillar dysentery, but this is not associated with hemorrhoids, as MacArthur points out: "The disease does not cause piles, people do not die of piles, and an epidemic of piles in any circumstances is to my mind incredible." MacArthur, W. P. "The Occurrence of the Rat in Early Europe. The Plague of the Philistines (1 Samuel, 5, 6)." *Transactions of the Royal Society of Tropical Medicine and Hygiene* (1952) 46:210–11, quoted in Marks, Geoffrey, and Beatty, William K. 1976. *Epidemics*. New York: Charles Scribner's Sons, p. 9.

3. Hirst, p. 8.

4. As Hirst puts it, "In all probability the 'mice' in the Philistine towns and ships were really rats, probably *Rattus rattus alexandrinus*. Professor Haas found a skeleton indistinguishable from *R. rattus* in a neolithic site at Mount Carmel, Palestine. It is interesting to note the presence of rats in the passage, though it doesn't seem likely, from the simple mention of the 'mice that ravage your fields,' that the ancient Israelites themselves made the direct connection between the disease and the rodents."

5. In 430 B.C.E.; it lasted in Athens for over two years.

6. Thucydides. 1989. *The Peloponnesian War. The Complete Hobbes Translation*. David Grene, ed. Chicago: University of Chicago Press, pp. 116–17.

7. Simpson, W. J. 1905. *A Treatise on Plague Dealing with the Historical, Epidemiological, Clinical, Therapeutic and Preventive Aspects of the Disease*. Cambridge: Cambridge University Press, p. 4. Wu Lien-teh also considers this account an "undoubted trace of plague": 1936. "Historical Aspects," in Wu, L-t, Chun, J. W. H., Pollitzer, R., and Wu, C. Y. *Plague: A Manual for Medical and Public Health Workers*. Shanghai: Mercury Press, p. 2.

8. Simpson, W. J. 1905. *A Treatise on Plague Dealing with the Historical, Epidemiological, Clinical, Therapeutic, and Preventative Aspects of the Disease.* Cambridge: Cambridge University Press, pp. 4-5.

9. Dante, John Ciardi's translation, quoted in Evans, J. A. S., *The Age of Justinian.* London and New York: Routledge & Kegan Paul, p. 9. Evans points out that Dante places Justinian in Paradise, but he also cites Ciardi's comment that "another reading of history might suggest several pits that might have claimed Justinian."

10. Gibbon, Edward. *The Decline and Fall of the Roman Empire.* Vol. 2. New York: Modern Library, p. 481.

11. Treadgold, Warren. 1997. *A History of the Byzantine State and Society.* Stanford: Stanford University Press, p. 179.

12. As Gibbon says (p. 468), Theodoric lived too long: he lived to imprison and execute Boethius, perhaps the greatest philosopher of late antiquity, for suspected treasonable contact with Justinian. While in prison, Boethius wrote his great work *The Consolation of Philosophy*, for which he is still remembered.

13. According to Procopius's *The Secret History*, G. A. Williamson, trans. London and New York: Penguin, 1982.

14. Thompson, James Westfall. 1928. *Economic and Social History of the Middle Ages.* New York and London: D. Appleton-Century.

15. Constantinople had had two ports, but the emperor Julian had built a third, on the Sea of Marmara, and Theodosios still another.

16. Thompson, p. 157.

17. Evans, p. 32.

18. Ibid., p. 34.

19. Ibid.

20. Wu C. Y. 1936. "Insect Vectors," in Wu et al., 1936, p. 260.

21. Pollitzer, Robert. 1954. *Plague.* Geneva: World Health Organization. "Most observers are agreed that the transport of infected fleas in goods . . . is of great importance in the spread of plague" (p. 387).

22. Procopius. *History of the Wars.* Vol. 2. H. B. Dewing, trans. Cambridge, MA: Loeb Classical Library, p. 453. All subsequent citations are from pp. 453–71.

23. Mee, C. 1990. "How a Mysterious Disease Laid Low Europe's Masses." *Smithsonian* 20 (Feb):66–79. Also: see, for example, Perry, R. D.,

and J. D. Featherstone. "*Yersinia pestis*—Etiologic Agent of Plague." *Clinical Microbiology Review* 10:35–66; and McGovern, T. W., and A. Friedlander, 1997. In R. Zajtchuk and R. F. Bellamy, eds. *Medical Aspects of Chemical and Biological Warfare.* Bethesda, MD: Office of the Surgeon General, pp. 2070–78.

24. David Dennis, personal communication with the author.

25. Biraben, J.-N., and Le Goff, Jacques. 1975. "The Plague in the Early Middle Ages." In Forster and Orest Ranum, eds., *Biology of Man in History.* Baltimore: Johns Hopkins University Press, p. 55.

26. Chun, J. W. H. 1936. "Clinical Features," in Wu et al., 1936, p. 313.

27. Historian Timothy Bratton, by estimating population densities, calculates the population of Byzantium during Justinian's day as about 288,300 people; he maintains that the oft-quoted figure of "hundreds of thousands" dead of plague in Byzantium alone (figures that derive from Procopius and John of Ephesus) "cannot be taken seriously." Bratton, Timothy, 1981. "The Identity of the Plague of Justinian, Part Two," in *Transactions and Studies of the College of Physicians of Philadelphia* 3(3):174–80.

28. Whitby, Michael. 2000. *The Ecclesiastical History of Evagrius Scholasticus.* Liverpool: Liverpool University Press, pp. 315–16.

29. Ibid., pp. 229–32.

30. Dalrymple, W. 1997. *From the Holy Mountain.* New York: Henry Holt, pp. 58–62.

31. Whitby, p. xiv.

32. This and following brackets and parentheses in the original. Harrak, Amir. 1999. *The Chronicles of Zuqnin, Parts III and IV.* Toronto: Pontifical Institute of Medieval Studies, pp. 94–95. Subsequent citations are from pp. 95–107.

33. Gregory of Tours. 1974. *The History of the Franks,* trans. Lewis Thorpe. London: Penguin, pp. 509–11. Subsequent citations are from pp. 510–11, 543–46.

34. L. Fabian Hirst puts it this way: "The account given by Gregory of Tours (540–94) of the importation of plague from Spain into Marseilles harbor in A.D. 588 in a trading ship is very strongly suggestive of rat-borne infection. In fact Sticker selected this epidemic from all those recorded by him in the whole course of history as the very type of an outbreak of rat origin. . . . As Sticker points out, a latent period between the introduction of infec-

tion into a locality and the flaring up of an epidemic is characteristic of plague of rat origin. During this period only a few sporadic cases of human plague occur, but the infection is being established in colonies of rats. Only when the rat epizootic is in full career does the human epidemic follow." Hirst, p. 125.

35. Paul the Deacon. 1974. *History of the Lombards*, trans., William Dudley Foulke; ed, Edward Peters. Philadelphia: University of Pennsylvania Press.

36. Thucydides, pp. 115–18.

37. Except for the ambiguous passages in Evagrius already discussed.

38. Pollitzer, R. 1954. *Plague*. Geneva: World Health Organization, p. 12; Simpson, 1905. *A Treatise on Plague*. Cambridge: Cambridge University Press, p. 16; Wu et al. 1936. *Plague: A Manual for Medical and Public Health Workers*. Weishen Shu, China: National Quarantine Service, p. 2.

39. De Voragine, Jacobus. 1941. *The Golden Legend*. Granger Ryan and Helmut Ripperger, trans. New York: Arno, p. vii.

40. Hirst, 1953, p. 110.

41. It was on the Nile grass rat, Gage believes, that *Xenopsylla cheopis* evolved.

42. Wu Lien-teh, the great Chinese-Malayan plague researcher of the Third Pandemic, thought a great deal about this issue and came, in 1936, to much the same conclusion as Gage: "we now know a whole series of endemic [African] plague foci, usually with epizootics among the wild rodents, situated near to or even contiguous with Central Asia. Indeed our studies have convinced us that the whole of this vast territory with its hosts of wild rodents might be compared with a heap of embers where plague smoulders continuously and from which sparks of infection may dart out now and then in various directions, later either to become extinguished or to set up conflagrations or slowly burning endemics according to the suitability of the fuel they find" (Wu Lien-teh, 1934, 1936).

43. It is a curious fact that people tend to forget horrible disease outbreaks. Until recent attention was focused on it, the terrible flu pandemic of 1918 had largely slipped from the popular consciousness; so had its appalling sequel, the outbreak of *Encephalitis lethargica* chronicled in Oliver Sacks's remarkable book *Awakenings*, and in the movie of the same name.

44. There are many contemporary sources examined by Biraben and Le

Goff that mention the disease only in passing, and so are not considered in the present discussion. For a complete bibliographic reference, the interested reader should consult this article. Biraben and Le Goff.

45. Biraben and Le Goff, p. 59.

46. Ibid.

47. Naphy, William, and Spicer, Andrew. 2000. *The Black Death and the History of Plagues, 1345–1730*. Stroud, Gloucestershire and Charleston, South Carolina: Tempus, p. 22.

48. Biraben and Le Goff, p. 63.

49. Bratton, 1981, p. 180.

50. Biraben and Le Goff, p. 63.

51. Bratton, 1981, p. 180.

## IV. BLACK DEATH

1. The Second Pandemic did not become known as the Black Death for centuries after the outbreak. It was referred to as the Pestilence or the Great Dying. I use the term Black Death—never Black Plague, which is a solecism—because it has become standard.

2. There is some evidence that bubonic plague never entirely vanished from Europe in the years preceding the Black Death. The plague saint St. Roch was born in 1295 and died in 1327, twenty years before the Black Death. He was stricken by plague at about age twenty, in the French town of Aquapendente, which was suffering an outbreak at the time. He was left alone to die of the disease, but, tended by his faithful dog, he recovered. Credited with miraculous powers against plague, St. Roch is always portrayed with a bubo on his thigh. Furthermore, according to scholar Michael W. Dols and plague expert W. J. Simpson, bubonic plague afflicted the army of Frederick II (Barbarossa) around Rome in 1167, and later attacked Rome itself. Plague also was reported in France and Italy during the thirteenth century. Very little is known of these outbreaks, which may or may not have been true plague; there is no affirmative reason, however, to believe that bubonic plague must have vanished from the Mediterranean area after Justinian's Plague brought it into the local rat populations. Dols, Michael W.

1977. *The Black Death in the Middle East.* Princeton: Princeton University Press, p. 31, n. 37. See also Simpson, pp. 20–21.

3. Saunders, J. J. 1971. *The History of the Mongol Conquests.* Philadelphia: University of Pennsylvania Press, p. 49.

4. Ibid., p. 67.

5. Grousset, René. 1970. *The Empire of the Steppes: A History of Central Asia.* New Brunswick, New Jersey: Rutgers University Press, pp. 248–49. Chingis had a simple credo: "To cut my enemies to pieces, drive them before me, seize their possessions, witness the tears of those dear to them, and embrace their wives and daughters," he was quoted as saying by his advisor Rashid ad-Din.

6. McNeill, William H. 1976. *Plagues and Peoples.* Garden City, New York: Anchor Books/Doubleday, p. 151. McNeill recognized the role of the Mongols in the dissemination of the Black Death. But he had that role backward. He proposed that the restless movement of the Mongol hordes after Chingis allowed the marmots and susliks of the steppe to come into contact with rats from "the borderland between India, China, and Burma." McNeill suggests that Mongol raiders in China might have stuffed their saddlebags full of grain or "some other form of booty," which might have contained live, infected rats and fleas. Once introduced to the native rodent populations, plague could have slowly diffused across the entire steppe (p. 160). But almost certainly the movement of plague went the other way, from the marmots of Central Asia to the rats of China, and to the warriors of the Great Khan themselves.

7. Grousset, p. 252.

8. Ibid., p. 253.

9. No actual copy has been preserved; all that is known of the *yasaq* comes from accounts by observers. Saunders, p. 69.

10. You may not pollute water by washing in it, the Mongols believed. This, and the injunction against ritual slaughter of animals, put them in direct conflict with Muslims, notwithstanding their express policy of toleration.

11. Pollitzer, 1954, p.4.

12. Mark Wheelis, personal communication with the author.

13. Suntsov and Suntsova, pp. 645–47.

14. Or, more probably, there were several such instances, though of course we can never know.

15. This section, and the quotations from Ibn al-Wardi and others, are based on the important work by historian Michael W. Dols. Dols's 1977 work, *The Black Death in the Middle East*, is hard to locate, but it is indispensable for anyone interested in the wider (non-European) history of the Black Death.

16. Ibid., p. 40. Dols cautions that "there is a danger in following Ibn al-Wardi's account of the dissemination of the plague too closely; in some cases, historical accuracy may have been sacrificed for the benefit of literary conceits and rhyme."

17. Ibid., pp. 38–39., citing *Risalat an-naba* of Ibn al-Wardi.

18. Horses apparently are not susceptible to plague, though camels die of it.

19. Dols, pp. 40–41.

20. Ibid., p. 41.

21. Campbell, Anna Montgomery. 1931. *The Black Death and Men of Learning*. New York: Columbia University Press. According to Campbell, Ibn al-Khatib also recognized that individuals in plague-infested communities, over time, apparently acclimate to it and develop immunity, a fact long denied by plague experts, and which has been recognized only very recently by plague researchers in Madagascar.

22. Dols, p. 94; Campbell, pp. 26–27.

23. Dols, p. 48.

24. Present-day Iran and Iraq.

25. Dols, p. 45.

26. The Mongols were generally, and incorrectly, known as Tartars in the Middle Ages. They had conquered the *Tatars*, Chingis's ancient enemies; a Frenchman, seeing their ghastly appearance, named them Tartars, from Tartarus, the hell of the ancient Greeks.

27. Horrox, Rosemary, trans. 1999. Excerpt from *Historia de Morbo* by Gabriele de' Mussis, in *The Black Death*. Manchester: Manchester University Press, p. 17.

28. Gasquet, Francis Aidan. 1908. *The Black Death*. London: George Bell and Sons, p. 5, n. 2.

29. Wheelis, Mark. 2002. "Biological Warfare at the 1346 Siege of Kaffa." *Emerging Infectious Diseases* 8(9):971–75.

30. Horrox, p. 46.

31. L. Muisis, who was from Tournai, claims to have "hunted out the most reliable accounts and the most accurate information I could find concerning the mortality," which he specifically wrote down so that later generations could have knowledge of it. Ibid., p. 19.

32. Ibid., p. 20.

33. The bedding certainly could have been contaminated with spittle or with human fleas. But as Dr. David Dennis, a plague expert formerly with the CDC, who treated plague patients in Vietnam, notes, it is difficult to believe the plague could kill overnight. Perhaps, though, with a highly virulent strain (more virulent than the strains in Vietnam), people really could die that quickly. Furthermore, it is possible that some plague strains allow for people to be ambulatory longer—see Chapter 7.

34. "Everyone has a responsibility to keep some record of the disease," de' Mussis tells us, "and because I myself am from Piancenza I have been urged to write more about what happened there in 1348." Horrox, p. 21.

35. Ibid., p. 20.

36. Pollitzer, p. 439.

37. Wu Lien-teh, 1926. *A Treatise on Pneumonic Plague.* Geneva: League of Nations Health Organization, p. 185. Wu gives numerous examples of pneumonic plague giving rise to bubonic plague through direct contact with pneumonic plague patients. He notes that "not so much the existence as the rarity of such cases calls for comment" (p. 186). Of course, during the Black Death, parasitism with human fleas was undoubtedly widespread—so that such bubonic or septicemic infection must not have been rare at all.

38. This symptom was never mentioned in Justinian's Plague; neither is it reported today, on which more below.

39. Horrox, pp. 24–25.

40. See Wheelis, 2002.

41. Bartsocas, C. S. 1996. "Two Fourteenth Century Descriptions of the Black Death." *Journal of the History of Medicine and Allied Sciences* 21 (October), p. 396.

42. Gasquet, p. 12.

43. According to Philip Ziegler, it may have been such a death ship that

first brought the Black Death to Bergen, Norway. The story goes that one infected sailor aboard a wool ship bound from England infected all the others; the ship, manned now only by the dead, drifted about until it washed up on the shore near Bergen. The hapless local people boarded the foundered ship, and brought back the infection. Ziegler, Philip. 1969. *The Black Death*. Godalming, Surrey: Bramley Books, pp. 84–85.

44. De Smet. "Breve Chronicon clerici anonymi." *Recueil de Chroniques de Flanders*. Vol. III, pp. 14–15, quoted in Ziegler, p. 6.

45. See also Wheelis, 2002.

46. Horrox, pp. 248–49.

47. Boccaccio's generous treatment of Jews in his famous *Decameron* is startling in an era where Jews, accused of poisoning wells and deliberately spreading plague, were broken on the rack or locked up in their synagogues and burned by the thousands.

48. Michael of Piazza, quoted in Nohl, Johannes. 1926. *The Black Death: A Chronicle of the Plague*. C. H. Clarke, London: George Allen and Unwin, p. 18. Subsequent citations are from pp. 18–20.

49. See Chapter 3, the section on Gregory of Tours.

50. Gasquet, pp. 40-41.

51. Ibid. p. 41.

52. Ibid. pp.41-42.

53. Nohl, p. 20.

54. Ziegler, p. 36.

55. Or *Bubboni* in modern Italian; in other words, buboes. Musa, Mark, and Bondanella, Peter. 1977. *The Decameron: A New Translation*. New York and London: Norton. p. 4., n. 2.

56. These excerpts are taken from ibid., pp. 3-4.

57. V. S. Vashchenok. 1988. Leningrad: Nauka.

58. Blanc, Georges, and Baltazard, Marcel. 1945. "Recherches sur le mode de transmission naturelle de la peste bubonique et septicémique." *Archives de l'Institute Pasteur de* Maroc III. Cahier 5, pp. 173–348; Blanc, Georges, and Marcel Baltazard. 1942. "Rôle des ectoparasites humaines dans la transmission de la peste." *Extrait du Bulletin de* l'Académie du Médecine, Séance du 20 Octobre 1942, Tome 126, No. 28-29-30, p. 446.

59. In an oft-quoted paragraph, Boccaccio describes how foul, filthy rags from a plague patient were tossed into the street, and how two pigs, nuzzling

the rags, suddenly stiffened and fell down dead. Pigs have been shown not to catch plague, and, in any case, such an immediate reaction to a germ is absolutely incredible. Yet Boccaccio claims to have witnessed the scene with his own eyes.

60. Biraben, Jean-Noel. 1975. *Les hommes et la peste en France et dans les pays européens et méditerranéens. Tome I: La peste e dans l'histoire.* Paris: Mouton, p. 74.

61. Ibid., pp. 90–91.

62. Ibid., p. 33.

63. Campbell, p. 3.

64. It appears to be something of an urban legend that the Jews of medieval Europe did not die of plague. One sees this often claimed in the literature, most recently in Norman Cantor's 2001 *In the Wake of the Plague: The Black Death and the World It Made* (New York: Free Press), but his claim is unsupported by any cited evidence. The only other contemporary source I know that discusses Jews falling ill from plague (besides Gui de Chauliac and Pope Clement VI) is Konrad of Megensburg, who says that so many Jews died of plague that the Jewish community had to build a new cemetery.

65. Gail, Marzieh. 1965. *Avignon in Flower, 1309–1403.* Boston: Houghton Mifflin, p. 188.

66. Gasquet, pp. 44–45.

67. Ibid., pp. 45–49.

68. Ibid., p. 49.

69. Nohl, p. 182.

70. Cohn, Samuel K. 2002. *The Black Death Transformed: Disease and Culture in Early Renaissance Europe.* London and New York: Arnold and Oxford University Press, p. 147.

71. Gasquet, p. 53.

72. See p. 10, this chapter.

73. Gasquet, pp. 60-62.

74. Horrox's translation puts the number at five hundred; but Gasquet, following a M. Geraud, who edited de Venette's chronicle for the French Historical Society, suggests that the number was actually fifty and that, furthermore, fifty is in keeping with other manuscripts, which give the same number. Horrox, p. 57; Gasquet, p. 55.

75. Hecker, J. F. C. 1846. *The Epidemics of the Middle Ages.* B. G. Babington, trans. London: Sydenham Society, pp. 32–33.

76. Hecker, p. 29.

77. Seward, Desmond. 1999. *The Hundred Years War: The English in France, 1337–1453.* New York and London: Penguin, pp. 41–75.

78. Ziegler, p. 94.

79. In Twigg, Graham. *The Black Death: A Biological Reappraisal.* 1984. London: Batsford Academic and Education.

80. Ibid., p. 139.

81. Scott, Susan, and Christopher Duncan. 2000. *The Biology of Plagues: Evidence from Historical Populations.* Cambridge: Cambridge University Press.

82. A third team from Pennsylvania State University, led by biological anthropologist James Wood, has announced its intention to publish a monograph that will make the same claim. Meanwhile they have released a public statement which asserts that modern "yersinial plague" is a "classic zoonosis," which cannot spread as rapidly as the Black Death did. None of these scholars think that pneumonic transmission provides an alternative. As Wood's team puts it, "We have constructed mathematical models of pneumonic plague, and these indicate that it cannot support a widespread, rapidly moving epidemic like the Black Death. The pneumonic plague explanation for the Black Death is therefore untenable."

It's a lot of weight to put on mathematical models, which are only as good as the assumptions behind them.

There is much more at stake here than an academic quibble. If *Yersinia pestis* is not the agent of the Black Death, then we have much less to fear from weaponized plague germs. Plague would be an ordinary bacterial infection with an overblown reputation—and we could disregard it as a biowarfare threat. But the Soviet scientists who understood plague, and who made *Yersinia pestis* into their chief bacteriological weapon, knew better.

83. Coulton, G. G. 1928. *The Black Death.* New York: Robert M. McBride.

84. Shrewbury, p. 46.

85. Gasquet, p. 138–40, taken from Friar John Clyn's *Annals of Ireland* (Irish Archaeological Society, 1849).

86. Ziegler, p. 157.

87. Gasquet, p. 148. Gasquet quotes Professor Thorold Rogers, who

wrote, "I have no doubt that the principal place of burial for Oxford victims was at some part of the New College garden, for when Wykeham bought the site it appears to have been one which had been previously populous, but was deserted some thirty years before the plague and apparently made a burial ground by the survivors of the calamity." The present warden of New College, Professor Alan Ryan, says that New College Cloister, instead of the garden, was the site of the plague pit (personal communication with the author).

88. In 2000, the publication by a noted team of French molecular biologists claimed to have isolated DNA from several Yersinia pestis genes from tooth-pulp taken from corpses in a known Black Death cemetary in southern France. As they put it in their abstract to their article: "We believe that we can end the controversy: Medieval Black Death was plague." Raoult, D., Aboudharam, G., Crubezy, E., Larrouy, G., Ludes, B., and Drancourt, M., 2000.[66] Molecular Identification by "Suicide PCR" of Yersinia Pestis as the Agent of Medieval Black Death." *Proceedings of the National Academy of Sciences* 97(23): 12800-03.

But despite this striking result, the experiment did not end the controversy; scientists on both sides of the debate have questioned the team's results from a number of perspectives. Later research in England by Alan Cooper and his team at the Ancient Biomolecules Centre at Oxford University failed to duplicate these results, according to an article by Debora MacKenzie in September 11, 2003, edition of *New Scientist* magazine.

In any event, the identification of the plague germ as the agent of the Black Death does not stand or fall on these molecular analyses.

89. Scott and Duncan, pp. 81–109; Twigg, pp. 113–16.

90. Naphy and Spicer, p. 34.

91. Hecker, p. 30.

92. Hecker's *Schwarze Tod* was published in Germany in 1832; Cardinal Francis Aidan Gasquet's *The Black Death* first appeared in 1893.

93. Hecker, pp. 8–9.

94. Gasquet, pp. 8–9.

95. According to Mongolian scientists, there is also a tiny plasmid found in marmot strains and nowhere else—though such plasmids, according to Vladimir Motin, are not all that uncommon. Still, no one seems to know what effect that that cryptic—or hidden—plasmid might have on the plague germ.

96. Wu Lien-teh. 1926, p. 3. By "old recorders" Wu means medieval chroniclers; he is specifically referring to the Black Death.

97. Ibid.

98. Ibid.

99. Except in virgin soil epidemics that have not evolved a history of resistance, such as the reported 75 percent mortality of smallpox among American Indians when they were first exposed to it. See Fenn, 2001.

## V. THE RENAISSANCE PLAGUE

1. Ratsitorahina, M., et al. 2000. "Epidemiological and Diagnostic Aspects of the Outbreak of Pneumonic Plague in Madagascar." *Lancet* 355:111–13.

2. Migliani, R., et al. 2001. "Resurgence de la peste dans le district d'Ikongo a' Madagascar en 1998." *Bulletin Société de Pathologie Exotique* 94(2):115–18. Smaller numbers of inapparent infections, because the number of exposed people was much smaller, were also detected by Ecuadoran scientists together with CDC experts: Gabastou, J. M., et al. 2000. "An Outbreak of Plague Including Cases with Probable Pneumonic Infection, Ecuador, 1998." *Transactions of the Royal Society of Tropical Medicine and Hygiene* 94(4):387–91.

Furthermore, plague skeptics Susan Scott and Christopher Duncan admit that their entire mathematical analysis of plague spread, using a system called Reed Frost Dynamics, cannot work if there are subclinical or inapparent infections present, which they deny. This illustrates a chief problem of relying on mathematical models: they are only as good as their initial assumptions—which in this case are incorrect at the start. Scott and Duncan, p. 29.

3. In the case of Madagascar, the F1 antigen discussed at length in Chapter 2.

4. Cohn, Samuel K. 2002. *The Black Death Transformed: Disease and Culture in Renaissance Culture.* London and New York: Arnold/Oxford University Press, pp. 212–16.

5. According to French flea expert J. C. Beaucornu, Oriental rat fleas don't even survive through the winter in cold climates. "Diversité des puces

vectrices en fonction des foyers pesteux." Manuscrit. No. 196 3/PLS14. *Journée Pen Hommage á Paul-Louis Simond.* But flea expert Abdu Azad of the University of Maryland insists that *Xenopsylla cheopis* can indeed overwinter in cold northern climates, so long as they remain near their rat hosts. These fleas, explains Azad, are intermittent feeders—they must feed every four to six hours to survive. This frequent feeding keeps the flea warm enough to survive, especially in the microclimate of the rat's burrow. Azad also notes, however, that all fleas show some seasonal differences in their behavior; though *Pulex irritans* can survive cold northern winters, it will be much more active in the summertime, and we can expect *Pulex*-borne plague to spread much more rapidly in the summer than in the winter.

6. Also see Audoin-Rouzeau F. 1999. "The Black Rat (*Rattus rattus*) and the Plague in Ancient and Medieval Western Europe" (in French). *Bulletin Société de Pathologie Exotique* 422–26.

7. Russell, Josiah C. 1948. *British Medieval Population.* Albuquerque, New Mexico: University of New Mexico Press, quoted in Carpentier, Elisabeth. "The Plague as a Recurrent Phenomena," in Bowsky, William M., ed., *The Black Death: A Turning Point in History?* Malabar, Florida: Robert E. Krieger, p. 37.

8. Samuel Cohn, a historian who denies that the Black Death could have been caused by *Yersinia pestis*, cites this children's plague as additional evidence, since, he claims, true plague does not produce immunity. He is wrong. No one seems to know how long immunity to plague lasts. But that there is some immunity after infection, perhaps lasting for years, cannot be denied, or vaccination (with live EV strain) would not work. And it does. There would also be no evidence of subclinical or inapparent infections—and there are: see Migliani et al.; Gabastou et al.

9. Slack, Paul. 1985. *The Impact of Plague in Tudor and Stuart England.* Oxford: Clarendon Press, p. 11.

10. Tomalin, Claire, 2002. *Samuel Pepys: The Unequalled Self.* New York: Knopf, p. 166.

11. A search of Shakespeare turned up a bare handful of references to fleas, one of which appears as the epigraph to this chapter.

12. Dickson, James H., Oeggl, Klaus and Handley, Linda L. 2003. "The Iceman Reconsidered," *Scientific American* (May).

13. Slack, p. 68.

14. Ibid., p. 69.

15. Ibid., p. 105.

16. Blanc, Georges, and Baltazard, Marcel. 1945. "Recherches sur le mode de transmission naturelle de la peste bubonique and septicémique" ("Research on the Mode of Transmission of Bubonic and Septicemic Plague"). Archives de L'Institut Pasteur de Maroc, p. 342, author's translation.

17. Blanc and Baltazard insist that rat fleas find human beings repugnant and will bite them only under conditions of absolute necessity. When asked about this, plague flea researcher Joe Hinnebusch comments, "Well, they bite me."

18. Blanc and Baltazard, 1945, p. 342.

19. Slack, p. 147.

20. Naphy and Spicer, p. 114.

21. Ibid.

22. Bell, Walter George. 1994. *The Great Plague of London.* Facsimile edition of the original 1924 publication. London: Bracken Books, p. 8.

23. Naphy and Spicer, p. 113.

24. Bell, p. 9.

25. Ibid., p. 23.

26. Ibid., p. 39.

27. Shrewsbury, p. 467.

28. Noorthouk, J. 1773. *A New History of London, Including Westminster and Southwark.* Quoted in Shrewsbury, p. 446.

29. Bell, p. 107.

30. Ibid., p. 110.

31. Ibid.

32. Pepys, Samuel. 1894. *The Diary of Samnel Pepys,* ed., Henry B. Wheatley. London: George Bell & Sons, Vol. 4, p. 428.

33. Bell, p. 156.

34. Tomalin, 2002.

35. Pepys, Vol. 5, p. 46.

36. Pepys, Vol. 5, p. 78.

37. Pepys, Vol. 5, p. 66.

38. Bell, p. 140.

39. Pepys, Vol. 5, p. 65.

40. Bell insists this incident actually took place, as he found the account in the memoir of John Reresby, which was not printed until 1734, long after Defoe's own publication. Bell, p. 134.

41. There is another rat flea native to England, *Nosopsyllus fasciatus*, but it is not as good a vector as *cheopis*, and more important, it appears exceedingly reluctant to bite human beings, though it will in cases of starvation. It is much more cold-hardy than the Oriental rat flea *Xenopsylla cheopis*.

42. Blanc and Baltazard, 1945. 3(5):173–348.

43. Bell, pp. 142–43.

44. Ibid.

45. These figures are taken from Shrewsbury, pp. 456–59.

46. Bell, p. 136.

47. Cipolla, C. M. 1981. *Fighting the Plague in Seventeenth-Century Italy*. Madison: University of Wisconsin Press, p. 4.

48. Naphy and Spicer, p. 77.

49. Cipolla, 1981, p. 39.

50. Naphy and Spicer, p. 79.

51. Cipolla, 1981, p. 16.

52. Nuttall, Zelia. 1912. "An Historical Document Relating to the Prevention and Cure of Plague in Spain in 1600–1601." *Journal of Hygiene* 12(1):46–48.

53. Cipolla, 1981, p. 12.

54. Cipolla, 1981, pp. 12–13n.

55. Cipolla, Carlo M. 1979. *Faith, Reason, and the Plague: A Tuscan Story of the Seventeenth Century*, Brighton, Sussex: Harvester Press, p. 2.

56. Cipolla, Carlo M. 1973. *Cristofano and the Plague: A Study in the History of Public Health in the Age of Galileo*. London: Collins, p. 27.

57. Ibid., p. 83.

58. Kohn, George Childs. 2001. *Encyclopedia of Plague and Pestilence from Ancient Times to the Present*, revised edition. New York: Checkmark Books, pp. 172–3.

59. Ibid., p. 218.

60. Naphy and Spicer, p. 137.

61. Kohn, p. 219.

62. Naphy and Spicer, p. 143.

63. Alonso, J. M. 1999. "Interactions écologiques des Yersinia au sein de

l'hôte reservoir commun, le rongeur." *Bulletin Société Pathologie Exotique* 92(5.2):414-17.

64. Igor Domaradskij, personal communication with the author.

65. Another, older argument suggests that the gradual replacement of *Rattus rattus*, the black house rat, by the larger, fiercer Norway rat, *Rattus norvegicus*, which lives outdoors and not in houses, was responsible for the disappearance of plague. But this argument, too, is not compelling, and has been largely abandoned today. First, Norway rats invaded Western Europe only long after plague had already receded. Second, Norway rats carry plague as readily as the black ones do, and, though they live less intimately with humanity, nevertheless are also commensal rodents, and could also spread the plague. Pollizer, 1954.

In any event, it is probable that rats played only an introductory role in bringing plague into a human community, and that the dominant role in spreading the infection was the human flea. The Norway rat could have served as well as the black rat to move plague from one human neighborhood into another. And investigations have shown that during outbreaks in port cities in Egypt, Norway rats were the dominant species in the poor, plague-ridden residential quarters, while black rats were restricted to the docks. Pollitzer, p. 33.

66. Simpson, p. 350.

67. Kinglake, J. W. 1906. *Kinglake's Eothen.* London: Henry Frowde, pp. 6–7.

68. Ibid., p. 206.

69. Ibid, p. 180.

70. Simpson, p. 350.

71. Blanc and Baltazard, 1945.

## VI. THE THIRD PANDEMIC

1. According to flea expert Robert Elbel of the University of Utah, a large epidemic of this pestis minor or ambulatory plague occurred in Thailand in the 1950s; Elbel himself witnessed and reported on it.

2. Simpson, p. 161.

3. According to Russian plague expert Sergei Balakhonov, head of the Microbiology Department at the Anti-Plague Institute at Irkutsk, plague is endemic to the region, and, like plague among Siberian marmots, it is extremely virulent and dangerous to humans. Personal communication. The relationship between the Astrakhan pestis minor outbreak and the lethal Vetlianka epidemic remains unclear, but there certainly seems to have been a rodent epizootic preceding the epidemic (see Wu Lien-teh, 1926, pp. 128–29).

4. Even to Australia, though it did not remain there long, as there were no indigenous rodents to become a permanent reservoir.

5. In the past year alone, two popular books have examined different aspects of the story. One, Edward Marriott's *Plague*, deals largely with the competition between two Hong Kong scientists, Alexandre Yersin and Shibasaburo Kitasato, to discover the plague bacillus; the other, Marilyn Chase's *The Barbary Plague*, tells the story of plague and the plague fighters of San Francisco. Marriott, Edward. 2002. *Plague: a Story of Science, Rivalry, and the Scourge That Won't Go Away.* New York: Metropolitan Books; Chase, Marilyn. 2003. *The Barbary Plague: The Black Death in San Francisco.* New York: Random House.

6. Wu Lien-teh, 1936, p. 12.

7. Ibid., p. 13.

8. Ibid., p. 16.

9. Benedict, Carol. 1996. *Bubonic Plague in Nineteenth-Century China.* Stanford: Stanford University Press, p. 7.

10. Wu, 1936, pp. 18–19.

11. Ibid., p. 21.

12. Marriott. p. 6

13. Butler, Thomas. 1991 *Plague and Other Yersinia Infections.* New York: Plenum.

14. Butler goes on to say that it is clear that Kitasato, too, actually saw plague: he described the bacillus as clearly bipolar when stained with aniline dyes. But Kitasato shuffled and waffled enough in later descriptions to cast doubt on his own discovery. Butler is in no doubt that Yersin, despite the handicaps of his terrible working conditions, was the better bacteriologist. Ibid., p. 24.

15. These quotations are taken from translated excerpts from ibid., pp. 17-18.

16. Ibid., p. 25.

17. Ibid., p. 19. But as Hirst notes (p. 109), "[Yersin's] identification of the fly germs as those of plague were almost certainly as erroneous as Kitasato's claim to have found the plague bacillus in the blood of weeks-old convalescents." Hirst, however, does not explain why the guinea pigs injected with the fly infusion died of plague. He does concede a bit later that "there is experimental evidence that *Pasteurella pestis* [an early name for *Yersinia pestis*] may multiply in the guts of fleas and even kill them" (p. 160).

18. Gregg, Charles. *Plague: An Ancient Disease in the Twentieth Century.* Revised edition. Albuquerque: University of Mexico Press, p. 58.

19. Hirst, facing p. 160.

20. Ibid., pp. 152–53.

21. Ibid., pp. 158–59.

22. Gregg, pp. 60–61.

23. Hirst, p. 174.

24. Ibid., p. 138.

25. Pollitzer, 1954, pp. 484–85.

26. Hirst, p. 139.

27. Bacot, A. W., and C. J. Martin. 1914. "Observations on the Mechanism of Transmission of Fleas," *Journal of Hygiene* 13:423–39.

28. Hirst, p. 186. As Hirst explains, two other researchers, Eske and Haas, found in 1940 that *Xenopsylla cheopis* blocked more efficiently than any of the seventeen flea species they studied.

29. Ken Gage, personal communication with the author.

30. Hirst, pp. 220–21. Emphasis added.

31. Hirst, p. 189. Wu, in his 1926 *Treatise on Pneumonic Plague*, disputes this interpretation—he insists that the swollen axillary lymph glands are actually only fatty tissue.

32. Wu Lien-teh. 1959. *Plague Fighter: The Autobiography of a Modern Chinese Physician.* Cambridge: W. Heffer & Sons, p. 53. See also Farrar, Reginald. 1912. "Plague in Manchuria." *Proceedings of the Royal Society of Medicine.* Vol. 5, Part 2. London: Longmans, Green, p. 5.

33. Farrar, p. 5. While dogs sometimes show seropositivity, and are used

as sentinel animals in Madagascar to see if plague is active among rodents in an area, they have never been shown to suffer from plague.

34. Gray, J. Douglas. 1911. "Report on Plague Outbreak in Manchuria and North China." *Lancet,* 1(2): 1152. Gray also mentions earlier outbreaks, but these are from the Volga-Astrakhan area, and have a different rodent reservoir. Gray records outbreaks beginning in 1906; Simpson's *Treatise on Plague* has accounts of pneumonic plague in the general region going back as far as 1894 (p. 224).

35. Gray, p. 1154.

36. Simpson, p. 224.

37. Chernin, Eli. 1989. "Richard Pearson Strong and the Manchurian Epidemic of Pneumonic Plague, 1910–1911." *Journal of the History of Medicine and Allied Sciences* 44(3): 297.

38. Gray, p. 1154; Wu, in his 1959 autobiography, discusses relations among all the countries with an interest in Manchuria as part of his narrative, pp. 1–74. Flohr, Carsten. 1996. "*The Plague Fighter: Wu Lien-teh and the Beginning of the Chinese Public Health System.*" Annals of Science, pp. 367–70; Nathan, Carl F. 1967. *Plague Prevention and Politics in Manchuria,* 1910–1931. East Asian Research Center, Harvard University. Cambridge, Harvard University Press, pp. 1–41.

39. Wu, 1959.

40. Chernin, p. 299.

41. Farrar, p. 5.

42. Gray, p. 1156.

43. Flohr, p. 373.

44. Wu, 1959 p. 249.

45. Flohr, p. 379.

46. Wu, 1959, pp. 19–20.

47. Ibid., pp. 20–24.

48. R. Willis, quoted in Gray, p. 1158.

49. A few cases did actually appear in Beijing, but the epidemic never took root there. Wu, 1959, pp. 33–34.

50. Wu, 1959, p. 36–37; Nathan, p. 13.

51. Nathan, p. 11.

52. Wu Lien-teh, Chun, J. W. H., and Pollitzer, R. 1923. "The Role of the Tarabagan in the Epidemiology of Plague." *Journal of Hygiene* 21:329.

53. Wu Lien-teh, and Eberson, Frederick. 1917. "Transmission of Pulmonary and Septicaemic Plague Among Marmots," in *Journal of Hygiene*, 16(1):1.

54. Ibid., p. 7.

55. Zhu Jin-quin et al. 1993. "A Study of the Epidemic Patterns and Control Measures of Human Plague in Qinghai Province." *Endemic Diseases Bulletin* 8(1):1.

56. Personal communication with the author.

57. Farrar, p. 12.

58. Ibid., p. 8.

59. Wu Lien-teh, Chun, J. W. H., and Pollitzer, R. 1923. "Appendix." *Journal of Hygiene* 21:342–58.

60. Wu, 1959, p. 104.

61. Dr. Wu does not give a first name for this researcher, who came from the Rockefeller Foundation.

62. Wu, 1959, p. 106.

63. Ibid, p. 106–7.

64. Ibid, p. 110.

65. Ibid, pp. 111–12.

66. Nathan, pp. 67–68.

67. Supposedly three people are reliably thought to have survived from the two Manchurian epidemics combined.

68. Michael Surgalla was a highly regarded plague expert, concentrating on basic research, who ran the plague laboratory for years at Fort Detrick.

69. Personal communication with the author.

70. Ibid.

71. Wu Yu-lin, 1995. *Memories of Dr. Wu Lien-teh, Plague Fighter.* Singapore: World Scientific Publishing Company.

72. Nathan, p. 51.

73. Ibid., p. 73.

74. Ibid., pp. 72–73.

75. Zhu.

76. Ingelsby, Thomas M. 2001. "Bioterrorist Threats: What the Infectious Disease Community Should Know About Anthrax and Plague." *Emerging Infections* 5. W. M. Scheld, W. A. Craig, and J. M. Hughes, eds. Washington: ASM Press.

77. The only strains that seem less virulent for guinea pigs are the vole strains of the Caucasus.

78. Meyer, Karl F. 1961. "Pneumonic Plague." *Bacteriological Review* 25:249.

79. Igor Domaradskij, personal communication with the author.

## VII. THE ENDURING THREAT

1. Personal communication with the author.

2. Personal communication with the author, September 1999.

3. Harris, Sheldon, 1994. *Factories of Death: Japanese Biological Warfare 1932–45 and the American Cover-Up*. London & New York: Routledge, p. 97.

4. Ibid., p. 107.

5. Ibid., pp. 78–79, 110.

6. Domaradskij and Orent, p. 125.

7. Harris, p. 80.

8. Ibid., pp. 204–21.

9. Zelicoff, A.P. 2003. "An Epidemiological Analysis of the 1971 Smallpox Outbreak in Aralsk, Kazakhstan." *Critical Reviews in Microbiology* 29(2):97-108. Some American scientists have publicly speculated that the so-called Aralsk strain may have been genetically engineered for vaccine resistance. But the Soviet program did not even begin to consider genetic engineering until 1973, when Victor Zhdanov and Igor Domaradskij were put in charge of bioweapons design and development and, by focusing Soviet bioweapons efforts on producing altered strains, effected a revolution. Most scientists, including leading pox virologist Peter Jahrling of USAMRIID, believe that the Aralsk strain was either identical to or very like the India-1967 (or India 1, as it is sometimes called) strain used as a basis for the later Soviet smallpox weapon.

10. Domaradskij and Orent, p. 64.

11. Personal communication with the author.

12. These strips of DNA-encoded endorphins, brain protein fragments, conotoxin (a lethal poison from shellfish), interferon alpha, and pox virus sequences. They also encoded some of the interleukins, which are immune system chemicals that can be poisonous in great quantities. Research on the

Interleukin-2 gene continued even after Popov left Vector for Obolensk in 1986. In Popov's words, these were "certainly some of the biggest synthetic projects of that time." Even in the Soviet bioweapons world, "very few people knew about" this super-secret research (personal communication with the author).

13. Biopreparat, however, was rather skeptical. In its opinion, says Popov, "the effects of peptides were 'too mild.'"

14. Personal communication with the author.

15. Personal communication with the author.

16. Personal communication with the author.

17. Popov bases this on his reading of an open 1991 paper by Volkovoi, in which he described the expression of a different protein, a fibrolysin, by this method.

18. Smallpox, a virus, is the main *virological* threat in both U.S. and Russian eyes.

19. Personal communication with the author.

20. Inglesby.

21. See Leonard Cole, 2003, *The Anthrax Letters.* Washington D.C.: Joseph Henry Press, pp. 163–71, 214–17.

22. Personal communication with the author.

23. Druett, H.A. et al. 1956. "Studies on Respiratory Infection II. The Influence of Aerosol Particle Size on Infection of the Guinea-Pig with *Pasteurella Pestis.*" *Journal of Hygiene* 54:1.

24. Meyer, Karl F. "Pneumonic Plague." *Bacteriological Review* 1961; 25:249–61.

25. Wells, W. F. and M. W. Wells. 1936. "Airborne Infection." *Journal of the American Medical Association* 107:1698–1703; cited in Meyer.

26. Meyer, 1961, p. 253.

27. Igor Domaradskij, personal communication with the author.

28. Soviet bioweaponeers, Domaradskij says, *always* leave some window of vulnerability in their antibiotic-resistant strains—or the disease could not possibly be treated in the case of an accidental infection or a laboratory or bioweapons accident.

29. According to Sergei Popov's account, which he has from Ken Alibek, in 1979, a technician at the secret biological weapons laboratory located in the city of Sverdlovsk forgot to change a filter on an anthrax fermenter. The

remaining filters eventually broke, and a small explosion, caused by increased pressure in the exhaust line, allowed a small amount (probably only a few ounces) of virulent anthrax to escape into the atmosphere. At least sixty-eight people contracted anthrax and died. Some former Soviet officials even now continue to deny that this was a bioweapons accident. They claim that the people died of eating "infected meat." But the anthrax victims all died of inhalational, not intestinal anthrax. Says one American scientist, "They would have had to inhale their meat."

30. E-mail to the author.

31. This is from an e-mail Kislichkin wrote to me, after objecting to my account of his firm and his flyer in an article I wrote for *The American Prospect*. "After Anthrax," 11(12).

32. From Ingelsby et al. 2000, "Plague As a Biological Weapon: Medical and Public Health Management." *JAMA* 283(17):2281–90.

33. Galimand, M., et al. 1997. "Multidrug Resistance in *Yersinia pestis* Mediated by a Transferable Plasmid." *New England Journal of Medicine* 337(10):677–80.

# ACKNOWLEDGMENTS

Many people have contributed to this book; without their help the entire adventure would have been impossible. I am grateful to them all.

There are three people I must thank first. Without Michael Skube, who brought me into the writing life, I would never have become a writer. He has been a close friend and mentor since I first wrote for his book section in the *Atlanta Journal-Constitution.* Paul Ewald's intellectual influence permeates this book. We both trained with zoologists at the University of Michigan, Dr. Ewald as a post-doc, I as an anthropology-zoology honors student. Dr. Ewald went on to revolutionize the study of the evolution of infectious disease: evolutionary epidemiology, the subfield of biology that deals with the evolution of disease agents, is essentially his creation. He has patiently worked through with me all of the evolutionary ideas that appear in this book, steering me straight when he saw me veering off course. Igor Domaradskij has been a dear friend and a constant source of ideas and information. As one of the greatest living experts on *Yersinia pestis*, he has been an invaluable resource; he was also my introduction to the fascinating world of plague research. My family and I also appreciate the warm hospitality and friendship of his family: his wife, Svetlana, his daughters, Tatiana, Anna, and Anastasia, and his son-in-law, Nickolai, who drove us to Obolensk.

Domaradskij's friend Lev Melnikov and his wife, Lida, also offered us their kind hospitality; Dr. Melnikov shared some of his knowledge and his memories as a plague fighter with me. Robert

Brubaker of Michigan State University has had great patience for my many questions over the four years I have known him. Vladimir Motin, of the University of Texas Medical School, has been an endless source of inspiration, guidance, and generous criticism. Kenneth Gage, Chief of the Plague Section of the Centers for Disease Control and Prevention in Fort Collins, Colorado, has been another singular inspiration. Dr. Gage has given me a great deal of his time and knowledge over the four years since I first walked into his office. Many of the ideas in this book have been worked over extensively with him (which does not mean that he necessarily agrees with what I have written). I thank him also for his thoughtful criticism of this book.

I thank Michael Kosoy of the CDC and his wife, Olga, for their warmth and hospitality; Dr. Kosoy first made me think about the idea of co-evolution of virulence between the plague germ and its various host species, and was the first to introduce me to the concept of host-specific strains in plague. He also read and gave me helpful comments on several of my chapters. David Dennis, formerly of the CDC, has also been generous with his time and expertise; I greatly value his criticism of an earlier version of Chapter 3. Scott Bearden, Erlan Ramanculov, and May Chu also took the time to share some of their knowledge with me; to Dr. Bearden I owe a better understanding of those remarkable plague virulence factors called Yops. John Montineri spent the better part of a day driving me to see prairie dog colonies within and just outside Fort Collins, and also showed me a little of his world.

I thank Joseph Hinnebusch and Tom Schwan of Rocky Mountain Laboratories for explaining many aspects of flea plague evolution; Dr. Hinnebusch kindly read and criticized my chapter on plague biology.

Lester Little of Vassar discussed the Justinian Plague with me, and introduced me to the source material on John of Ephesus; Timothy Bratton of Jamestown College sent me his fascinating articles on that first pandemic. Mark Wheelis discussed the dynamics of the Black Death with me innumerable times, and graciously sent me references on the Manchurian outbreaks of the twentieth century. Alan P. Zelicoff explained aspects of immunology, and found the references to subclinical plague infections in the literature for me. I greatly value his friendship and encouragement.

Sergei Popov and Ken Alibek have been generous with their time and have helped me understand many aspects of the Soviet bioweapons program. Dr. Popov spent many hours with me, explaining what has been done and what could still be done to create new and more dangerous forms of plague and other diseases. His explications have been invaluable. Christopher Davis shared some of his considerable knowledge of the Soviet bioweapons program with me.

Richard D. Alexander, my undergraduate advisor and lifelong friend, taught me to think like an evolutionist in the first place; he thoughtfully criticized my writing on the evolution of disease.

I also thank Robert Perry, Sarah Richardson, Elisabeth Carniel, Abdu Azad, Bakyt Atshabar, Bakhtiar Suleimenov, Gary Anderson, Seth Carus, Andrew Weber, Peter Jahrling, Gerry Andrewes, Anne Harrington, Thomas Butler, Steven Wampler, Pat Fitch, Emilio Garcia, John Hoogland, Paul Keim, Elena Gold, William Patrick III, Dong Zheng Yu, V. Zavyalov, Sergei Balakhonov, Barbara Reynolds, Bradd Shore, Deborah Kapel, all of whom helped me in various ways, and at various stages of this project. Raymond Zilinskas started me off on this journey, sending me a copy of *Troublemaker*, an English translation of Domaradskij's Russian memoir, and putting me in contact with Igor Domaradskij in the first place.

John Wright, my wonderful, inimitable agent, has been a constant source of advice and encouragement; I cannot imagine having undertaken this project without him. I also had the good fortune to work with Bruce Nichols of the Free Press, whose editorial instincts I have learned to rely upon, and who helped me and my book in ways I could not have imagined.

My son, Jonathan, came to Russia with us and has been patient and understanding through all his mother's vicissitudes. Finally, I owe my greatest thanks to my husband, Mitchell Goodman, who has been an excellent critic, thoughtful interlocutor, and loving support through it all.

# INDEX

Page numbers in *italics* refer to illustrations.


Africa, 58, 90–91, 92, 94
Alibek, Ken, 17, 215, 217, 232
Alonzo, J. M., 167–68
Anderson, Gary, 59
Andropov, Yuri, 21
anthrax, 17, 29–30, 132, 181, 212, 214, 216, 219, 220, 221, 223, 224, 227, 233
antigens, 37, 38, 44–46, 47, 48, 144
Athenian plague, 65–66, 74, 77, 88
Atshabar, Bakyt, 46–47, 56–58, 59, 121, 196, 223
Attila the Hun, 99, 100, 101, 108

Bacot, A. W., 186
Balakhonov, Sergei, 45, 51
Baltazard, Marcel, 121, 149–50, 158, 170–72
Bearden, Scott, 45–46, 48
Bell, Walter George, 150, 151–52, 154, 159
Beria, Lavrenty, 19–20, 26
*Biohazard* (Alibek), 17
biological weapons, 211–33
  aerosol, 212, 214–15, 220–21, 225–27, 231
  antibiotics for, 20, 23, 207, 216–17, 230–31
  binary combinations in, 23–24
  British research on, 224–26, 232
  catapulted corpses as, 109–10
  genetic modifications in, 21–22, 215, 216–21, 223
  Japanese use of, 43, 212–14, 215, 224, 231
  laboratories for, 18–19, 21, 24, 36, 206–7, 215–20
  literature on, 10, 17, 22–23
  moral implications of, 14, 211–12
  plague strains used in, 5, 29–30, 36, 43, 58, 138, 207, 211–33
  proliferation of, 227–29
  safeguards against, 15, 24, 229, 230–33
  Soviet research on, 5, 9–30, 36, 58, 206–7, 211–12, 214–21, 224, 226–29, 231–32
  testing of, 214–15, 221
  threat of, 29–30, 207, 211–33
  U.S. research on, 15, 16, 20–21, 202–3, 211–12, 214, 218, 221–25
*Biology of Plagues* (Scott and Duncan), 132–33
Biraben, J.-N., 93–94, 95, 122
Black Death, 99–140
  altered forms of, 136–39
  as bubonic plague, 113, 115, 116–20, 121, 123, 124, 126, 132, 136–37, 139, 143
  climatic factors in, 106–7, 117, 147
  contagiousness of, 111, 113–22, 124, 125, 132–33, 135, 136–40
  contemporary accounts of, 90, 106–29, 133–35, 137, 138–39, 143
  corpse disposal in, 116, 126
  death toll from, 3, 94, 111, 114–21, 123, 126, 133, 135–40, 145, 177, 185–86
  in Europe, 111–35, 143, 145
  fleas as carriers of, 3, 121–22, 124, 126, 132, 133, 135, 138, 230
  geographic area of, 94, 107–14, 117–18, 143

Black Death *(cont.)*
humans as carriers of, 118, 121–22, 124, 126, 128–29, 138–40
Jews blamed for, 124–25, 128, 130
marmots as carriers of, 103, 105–6, 137–38, 139, 205, 206
medical assistance in, 24, 28, *98*, 111, 120, 123–24, 133
as *Medievalis* biovar, 58
origins of, 99–106
outbreaks of, 4, 103–14
as pandemic, 2, 3, 67, 111, 140, 145–46, 175
pestilential odor of, 137, 138–39
plague strain in, 38, 50, 57
as pneumonic plague, 107, 116–17, 122, 124, 125, 126–27, 133, 137–39, 140, 143, 222, 230
rats as carriers of, 104–5, 106, 112, 121, 124, 126, 132, 135, 230
recovery from, 143–44
religious explanations for, 89, 111, 119, 129–31
Renaissance Plague compared with, 143–44, 145, 147, 148, 164
as Second Pandemic, 67, 145–46, 175
as septicemic plague, 121, 122
social impact of, 114–21, 124–25, 129–31, 140, 233
successive waves of, 140, 143–46
symptoms of, 84, 113, 115–22, 123, 137, 139–40
Third Pandemic compared with, 176–77, 185–86, 194, 205, 206
trade routes and, 100–101, 102, 107–8, 111, 113–14, 131–32, 135
transmission of, 104–6, 107, 117–18, 132–33, 137–40
virulence of, 50, 90, 102, 105–6, 115–18, 120, 121–22, 132–33, 136–40, 221, 227, 230
warfare and, 108–11
*Black Death, The* (Nohl), 128
*Black Death, The* (Twigg), 132
Blanc, Georges, 121, 149–50, 158, 170–72
Boccaccio, Giovanni, 2, 114–15, 118–21, 122, 177
Bogan, Louise, 173
Bratton, Timothy, 95
Brezhnev, Leonid, 21
Brubaker, Robert, 22–23, 36–38, 45, 48, 202–3, 211, 222
bubonic plague:
antibiotics for, 35, 39
anti-immune mechanisms of, 35, 44, 77–78
biblical description of, 63–64, 88
Black Death as, 113, 115, 116–20, 121, 123, 124, 126, 132, 136–37, 139, 143
buboes formed in, 35, 44, 77–78, 79, 89, 90, *98*, 113, 115, 116, 117, 119–20, 121, 123, 126, 136, 143, 166, 184, 188, 225
fever of, 113, 116, 117, 123, 136, 166, 175
internal organs attacked by, 117, 125, 136, 137, 138–39
in nonhumans, 84, 90–93
pneumonic plague compared with, 34–35
Renaissance Plague and, 166–67, 171–72
symptoms of, 35, 44, 77–78, 79, 88–89, 90, 93, 116–22, 175
Third Pandemic as, 179, 184, 188, 196, 200–201, 225
transmission of, 34–35, 39, 44, 55, 57, 175, 230
Burgasov, P. N., 19–20
Burnazyan, A. I., 19
Byzantine Empire, 67–75, 83–85, 89–90, 93–96

Central Asia, 56, 57, 58, 59–60, 90–91, 99–107, 137–38, 140, 143, 167, 175, 176, 177, 179, 227
Chagatai, 103–4, 106, 107, 108
Chauliac, Gui de, 123–25, 126, 182, 232
China, 43, 54, 57, 90, 100, 103, *174*, 175, 176, 177–80, 187–205, 212–14, 222, 232

Chingis Khan, 99–104, 106, 107
cholera, 19, 214
Christie, Dugald, 195
Chu, May, 37–38
*Chuma* (Domaradskij), 10, 12
Chun, J. W. H., 202
Chwolson, Daniel Abramovich, 104
Clement VI, Pope, 123–25, 126, 130, 136
Clyn, John, 133–35

*Decameron* (Boccaccio), 2, 118–22
de Covino, Simon, 117, 139
Defoe, Daniel, 158
de' Mussis, Gabriele, 109–14
Devignat, R., 58–59
Di Bortoli, V., 55–56
Dickie, Walter M., 54–55
diphtheria, 145, 220–21
*Doctor Faustus* (Marlowe), 7
Dols, Michael, 106, 107
Domaradskij, Igor V., 9–30, 51, 105, 110, 211, 214, 215, 216, 217, 218, 219, 220, 224, 227–28, 229, 232
Duncan, Christopher, 132–33

Eberson, Frederick, 203–4
Ebola virus, 4, 66, 217
*E. coli*, 20, 37, 231
Ecuador, 34, 43, 122, 144
Edward III, King of England, 131
Egypt, 91, 92, 93, 169
Ekfelt, Dr., 199–200
*Eothen* (Kinglake), 168–69
epizootics, 50, 51, 185
Evagrius Scholasticus, 76, 80–82, 90, 93
evolution, theory of, 4, 24–25, 42, 45, 57, 215, 218

F1 antigen, 44–46, 47
Ferber, D. M., 22–23
fleas:
  blocked, *32*, 40, 42–47, 53–54, 121, 184, 186
  human, 43, 81, 121–22, 124, 126, 133, 138, 144–47, 149–50, 158–59, 162–63, 167, 186, 196–97
  plague transmitted by, 3, 27, 28, 34–35, 39–47, 49, 53–54, 67, 73–74, 81, 85–86, 90, 91, 92, 93, 112, 121–22, 124, 126, 132, 133, 135, 138, 145–47, 149–50, 151, 152, 154, 158–59, 162–63, 167, 170–71, 176, 184, 185, 186, 196–97, 213, 214, 230
  rat, *32*, 40–45, 73–74, 91, 92, 121, 126, 132, 135, 138, 144–45, 147, 149, 158, 163, 170–71, 184, 185, 186, 213
  unblocked, 43–44, 46–47, 121
Flecker, James Elroy, 97
Foege, William, 21
France, 85–86, 122–29, 131, 164, 165–67
Franklin, J., *142*
Frost, Robert, 31
Fulco della Croce, 111

Gage, Kenneth L., 33–34, 42, 43, 54, 91–92, 121–22, 205
Gasquet, Aidan, 137
gerbils, 27, 28, 51, 56, 57, 58, 91, 93, 103, 104, 205, 222–23
Germany, 125, 126, 129–31
Gibbon, Edward, 69, 75
Gilles Li Muisis, 110–11, 128–29
*Grande Chirurgie* (Gui de Chauliac), 123–25, 126
Great Britain, 131–33, 135, 145–46, 147, 148–60, 161, 163–64, 167, 224–26
Great Plague of London (1665), 148–60, 164
*Great Plague of London, The* (Bell), 150
Gregory I, Pope, *62*, 86–87, 89, 131
Gregory of Tours, 85–87, 89, 90
Grousset, René, 101, 102

Haffkine, Paul, 191–92, 193
Haffkine, Waldemar, 191–92
Hecker, J. F. C., 75, 136–37
hemin storage locus (HMS) gene, 42
*Henry IV, Part I* (Shakespeare), 141
Herodotus, 99

Hinnebusch, Joe, 40–41
*History of the Wars* (Procopius), 75–76
Hong Kong, 179–80, 181, 184
Hoogland, John, 52–53, 54, 56
Hoogland, Margaret, 54

Ibn al-Khatib, 107
Ibn al-Khatimah, 107
Ibn al-Wardi, 106
immune system, 35, 39, 41–50, 77–78, 144, 217–18
India, 2, 90, 175, 177, 182–83, 184, 185, 186–87, 191–92, 204–5, 230
influenza, 4, 55–56
Inglesby, Thomas, 222
Interleukin-10 (IL-10), 48–49
Ireland, 133–35
Ishii, Shiro, 212–14, 224, 231
Italy, 86–87, 89, 108, 111–19, 121, 122, 131, 160–64

Jackson, Jane, 54
Jahrling, Peter B., 212
Japan, 43, 204, 212–14, 215, 224, 231
John Cantacuzene, Emperor of Byzantium, 113
John of Ephesus, 61, 82–85, 90
Justinian I, Emperor of Byzantium, 67–75, 79–80, 95–96, 99
Justinian Plague, 67–96
  Black Death compared with, 75, 84, 89, 90, 94, 99, 102, 104–5, 106, 111, 117–18, 122, 123, 135
  contemporary accounts of, 74–89, 90, 93, 111, 135
  corpse disposal in, 79–80, 83–85
  death toll from, 75, 79–81, 83–85, 94, 117–18
  as First Pandemic, 67, 99, 135
  fleas as carriers of, 73–74, 81, 85–86, 90, 91, 92, 93
  grain storage and, 71, 73–74, 75, 76, 89–90
  health measures for, 79–81
  origins of, 73–74, 76, 77, 81, 89–93, 99
  outbreaks of, 73–74, 85–86, 89–93
  as pandemic, 4, 58, 67–96
  rats as carriers of, 73–74, 76, 85–86, 89–90, 91, 92–93, 102, 104–5, 106, 117, 118
  religious explanations for, 80–83, 86–87, 89
  social impact of, 74, 79–80, 87–88, 91, 93–96, 233
  successive waves of, 85–86, 89–90, 93–94, 102
  symptoms of, 77–79, 86, 90, 93
  Third Pandemic compared with, 78, 90, 176–77
  trade and, 74, 75, 85, 89–90, 92, 111, 117, 148
  transmission of, 73–74, 85–86, 89–90, 92–93, 104–5

Kaffa blockade (1345), 108–11, 113, 114
Kazakhstan, 46, 56, 101, 223, 230–31
Keim, Paul, 221
Khrushchev, Nikita, 24
Kinglake, A. W., 168–69
Kislichkin, N., 229
Kitasato, Shibasaburo, 181
Koch, Robert, 180, 181
Kosoy, Michael, 211–12
Kurdistan, 43–44, 58, 121

legionella, 216, 220
Le Goff, Jacques, 93–94, 95
Lewis, Frank, 199
lymph nodes, 30, 35, 44, *210*, 217, 226
Lysenko, Trofim, 24–25, 215

McNeill, William H., 102
Manchuria, 2, 33, 57, 138, 143, 177, 178, 187–205, 212–13, 222, 232
Marlowe, Christopher, 7
marmots, 11, 51, 56–60, 103, 105–6, 137–38, 139, 147, 175, 176, 177, 187–90, 195–96, 205, 206–7, 222–23, 227
Martin, C. J., 186
Martinevsky, I. L., 59
measles, 66, 74, 139, 145

Melnikov, Lev, 25–29, 30, 33, 36
Mesny, Dr., 192–93, 198
Meyer, Karl F., 205–6, 225, 226
Michael of Piazza, 115–16
Mongolian Empire, 99–111, 112
Motin, Vladimir, 37, 48, 49–50, 211, 219

Netesov, Sergei, 217–18
Newton, Isaac, 151
Nixon, Richard M., 20–21, 212

Ockham, William of, 135
*Okhotnik* (Hunter) program, 217–18

Pasechnik, Vladimir, 13, 17
Pasteur, Louis, 180
Patrick, William, III, 224
Paul the Deacon, 87–89, 93, 135
Pelagius, Pope, 86
*Peloponnesian Wars, The* (Thucydides), 65–66
peptides, 21–22, 216–17
Pepys, Samuel, 146, 155–57
Perry, Robert, 42, 48
*pestis inguinaria* (Oriental Plague), 117, 136–37, 139
*Pestoides* strain, 58–59
Petrarch, 114–15, 118
Philip VI, King of France, 131
pits, plague, *142*, 153
plague:
adaptation by, 57–58, 136–39
anthrax compared with, 29–30, 221, 224, 227, 233
antibiotics for, 2, 20, 23, 35, 39, 54, 143, 182, 198, 205, 207, 216–21, 227, 232, 233
anti-immune mechanisms of, 35, 39, 41–50, 77–78, 144, 217–18
bacterium of, 18–22, 29, 35, 39–40, 43–46, 51, 139–40, 144, 180–83, *210*, 216–21, 223
biological process of, 38–50, 58–59, 132–33
British research on, 224–26
in children, 144, 145, 157
chromosome of, 20, 37, 42, 47, 59, 216–17, 219
climate and, 92–93, 147
demic vs. zootic, 112
diagnosis of, 18–19, 48–49, 55–56, 139–40, 143–44, 145, 165, 206
disappearance of, 5, 52–53, 143–46, 160, 164, 167–72, 175
ecology of, 50–52, 57, 92–93, 147, 194, 197–98
evolution of, 4–5, 37–38, 41, 45, 138–39
fever of, 1–2, 39, 165, 166, 191
field studies for, 52–54, 56–58
foci of, 3, 104, 147, 168, 171, 222–23
French research on, 51, 58–59
genetic modification of, 21–22, 44, 45, 216–21, 223
hosts for, 3–5, 27, 28, 34–35, 41, 46–50, 56–60, 139
human vulnerability to, 3–5, 38–50, 52, 59, 104, 108–14, 117, 124, 137–38, 171–72, 175, 176, 205–7, 221, 227–29
injector system of, 47–48
internal organs attacked by, 44, 45–46, 57, 60, 136, 196, 198, 206
literature on, 22–23, 63–67, 112, 202
in nonhumans, 3, 11, 23, 27, 28, 34–35, 38, 39–40, 50–60, 84, 90–93, 104–6, 187–90, 222–23
outbreaks of, 1–3, 5, 26, 27–29, 33, 38–39, 51–59, 170–72, 230–33
phages of, 18–19
plasmids of, 20, 22–23, 37, 47, 217–21, 231
prevention of, 4–5, 11, 18, 28–29, 51, 55, 170–72, 222–23, 230–33
reservoirs of, 3, 27, 50–52, 56, 91, 104–5, 139, 143–44, 175, 222–23, 226
resilience of, 194, 197–98, 203, 222–24, 233
resistance to, 51, 56, 57–58, 105, 143–44
scientific name of, 1, 183
smallpox compared with, 3, 38, 39

Soviet outbreaks of, 27–29, 33, 36–37, 198
Soviet research on, 5, 9–30, 33, 36–37, 45–50, 56–60, 105, 138, 196, 206–7, 222–23, 226–29
strains of, 4, 20, 21–22, 23, 33–38, 45, 50, 56–60, 90, 105, 132–33, 137–38, 167–68, 175–77, 205, 216–17, 230–33
symptoms of, 1–2, 38–50, 77, 143–44, 145, 165, 166, 191, 206
transmission of, 3, 4–5, 30, 38–50, 52, 74, 75, 89–90, 92, 104–6, 220–23
tropism of, 57, 138, 196
undiagnosed, 48–49, 139–40, 143–44, 145, 206
U.S. outbreaks of, 1–2, 34, 36, 38, 50, 52–56, 58, 205
U.S. research on, 20–23, 33–38, 46, 59, 202–3, 205, 222–23
vaccines for, 20, 21, 36, 144, 192–93, 221, 229, 232
vectors of, 39–40, 41, 46–47, 92, 121, 143–44, 186
virulence of, 2, 3, 4–5, 21–23, 35–36, 37, 45–46, 47, 49–50, 55–58, 203, 205–7, 220–27, 233
weaponization of, 5, 29–30, 36, 43, 58, 138, 207, 211–33
*see also* Black Death; Justinian Plague; Renaissance Plague; Third Pandemic; *specific types*
*Plagues and People* (McNeill), 102
pneumonic plague:
antibiotics for, 36, 216–17
Black Death as, 107, 116–17, 122, 124, 125, 126–27, 133, 137–39, 140, 143, 222, 230
bloody sputum from, 113, 115, 121, 123, 125, 133, 134, 136, 139, 143, 147, 191, 197, 205–6, 223, 225–27
bubonic plague compared with, 34–35
death toll from, 27, 33, 39
genetic modification and, 216–17, 231
in nonhumans, 53, 137–38
outbreaks of, 27–29, 33, 34, 36, 55–56, 230
quarantine for, 28–29, 36
secondary, 35–36, 56
symptoms of, 81, 116–17
Third Pandemic as, 176, 177, 186–98, 200, 204–6, 222, 225
transmission of, 27, 28, 34, 35–36, 39, 46, 53, 57, 205–6, 224–27
weaponization of, 58, 212–13, 216–17, 224–27, 231
Pollitzer, Robert, 112, 201–3, 211
Poore's Plague, 145–46
Popov, Sergei, 14–15, 215–20, 223, 228
prairie dogs, 40, 43, 50, 52–54, 56, 92, 105, 137–38, 223
Procopius, 74–79, 82, 85, 111, 118, 135
*Pulex irritans*, *see* fleas, human

rats:
black, 64, 76, 90–91, 104–5, 147–48, 170–71, 186
domestic vs. wild, 91, 92, 93, 104–5, 117, 145, 148, 168, 186
plague carried by, 50–51, 52, 54, 55, 56, 57, 58, 60, 64, 73–74, 75, 76, 85–86, 89–90, 91, 92–93, 102, 104–5, 106, 112, 117, 118, 121, 124, 126, 132, 135, 144–48, 149, 151, 158, 163, 167, 168, 170–71, 176, 177–86, 196, 204, 205, 230–31
territory of, 76, 86, 89–90
Raymond-Roger, Viscomte of Carcassonne and Beziers, 127–28
Renaissance Plague, 143–72
Black Death compared with, 143–44, 145, 147, 148, 164
bubonic form of, 166–67, 171–72
contagiousness of, 147, 154–55, 162
contemporary accounts of, 145–46, 153–57, 162–63
corpse disposal in, *142*, 153, 155, 166
death toll from, 151, 155–58, 159, 164, 166–67, 232

fleas as carriers of, 43, 92, 145–47, 149–50, 151, 152, 154, 158–59, 162–63, 167, 170–71
geographic patterns of, 147, 148–52, 161, 164, 165, 175
humans as carriers of, 144–47, 149–50, 162–63, 170–72
infected materials in, 148, 149, 161, 164, 165–66, 168
medical assistance in, 28, 150, 154, 156, 160, 162, 165–66
origins of, 140, 147–48, 150
as pandemic, 4, 144–46, 150, 159, 164, 170–72
pesthouses in, 43, 151, 154–55, 157, 161, 162–63, 165
poverty and, 146–52, 157, 159
public health measures in, 150, 151, 159–64, 168–70
quarantine in, 28, 150, 152–55, 159–60, 161, 162, 163–66, 168–69, 171
rats as carriers of, 144–48, 149, 151, 158, 163, 167, 168, 170–71
resistance to, 145, 163, 167–72
septicemic form of, 167
social impact of, 145, 153–57, 166–67
successive waves of, 148, 160
symptoms of, 147, 161, 165
trade routes in, 92, 148, 150, 165–67
transmission of, 144–45, 149–51, 158–60, 161, 165–72
Rolle, Richard, 135
Roux, Emile, 180, 181
Rufus of Ephesus, 66–67, 74

San Bonaventura, Antero Maria da, 162–63
Sandakhchiev, Lev, 14, 216, 217
Scott, Susan, 132–33
*Secret History* (Procopius), 75
septicemic plague:
Black Death as, 121, 122
fever of, 79
Renaissance plague as, 167
Third Pandemic as, 200–201, 225, 226
treatment of, 39
virulence of, 39, 115
*Seuche, Die* (Weber), *8*
Shakespeare, William, 141, 209
Shi Tao-nan, 178
Shrewsbury, J. F. D., 152
Simond, Paul-Louis, 177, 183–84, 185
Simpson, W. J., 169–70, 184
smallpox, 3, 9, 14, 21, 38, 39, 66, 74, 139, 145, 214–15, 217, 221, 227, 228
Spain, 161–62, 164
squirrels, ground, 50, 51, 54–56, 58, 92
Stalin, Joseph, 11, 24, 25–26
Strong, Richard Pearson, 195, 197
Suleimenov, Bakhtiar, 46–47, 56–58, 59, 121, 138, 196, 223
Surgalla, Mike, 202–3
susliks, 51, 57, 58, 103, 196, 222–23
Symeon the Younger, Saint, 81–82
Sze, Alfred Saoke, 190–91, 193

tarabagans, 1, 105, 187–90, 195–96
Teague, Oscar, 197
*Tempest, The* (Shakespeare), 209
Theodora, Empress of Byzantium, 68–69, 70, 72, 95
Third Pandemic, 175–207
altered forms of, 176, 177–78, 196
antibiotics for, 182, 198, 205, 207
autopsies in, 191, 199–200
bacterium strain in, 177, 180–83, 197–98, 206
Black Death compared with, 176–77, 185–86, 194, 205, 206
bubonic form of, 179, 184, 188, 196, 200–201, 225
contagiousness of, 189–90, 195–98, 200–201
contemporary accounts of, 78, 178–79, 186–87
death toll from, 2, 175, 176, 177, 179–80, 185–86, 192, 194–95, 200, 201
fleas as carriers of, 176, 184, 185, 186, 196–97
geographic range of, 175–76, 184–85

Third Pandemic *(cont)*
humans as carriers of, 176, 180, 185, 196–98
marmots as carriers of, 175, 176, 177, 187–90, 195–96, 205, 206–7, 222
medical assistance in, 33, *174*, 190–93, 198
as *Orientalis* biovar, 58, 59, 175–78, 204
origins of, 175–80
outbreaks of, 4, 90, 175–80, 182, 184–87, 198–201, 225
as pandemic, 58, 67, 90, 222
as pneumonic plague, 176, 177, 186–98, 200, 204–6, 222, 225
protective measures in, 33, 192–93, 198, 200
public health measures in, 180, 190–95, 199, 201, 203–5, 232
quarantine in, 193, 194–95, 200–201, 232
rats as carriers of, 176, 177–86, 196, 204, 205
reservoirs of, 187–90, 195–96
scientific research on, 176–77, 180–87, 190–93, 195–96, 222
septicemic form of, 200–201, 225, 226
symptoms of, 179, 184, 191
transmission of, 179, 183–84, 188–90, 195–201, 222
vaccinations in, 192–93
Thucydides, 65–66, 77, 88
*Treatise on Plague, A* (Simpson), 169–170
*Treatise on Pneumonic Plague, A* (Wu), 112
*Très Riches Heures, Les, 62*
tularemia (rabbit fever), 23–24, 216, 229
Turkmenistan, 27–29, 101
Twigg, Graham, 132, 137, 145
typhus, 171, 214

Urakov, Nickolai, 16, 23–24, 25, 58, 217, 218, 220, 227

Vantigen, 37, 48
Venezuelan equine encephalomyelitis (VEE), 217–18
Vietnam, 57, 181–82, 196, 204, 230
Vincent, Thomas, 154
viruses, 4, 21–22, 44, 66, 216–19
voles, 58–59
Volkovoi, K. I., 220–21

Weber, Andrew, 212
Weber, A. Paul, *8*
Wheelis, Mark, 104
World Health Organization (WHO), 34, 164, 202, 211, 230, 231
Wu Lien-teh, 112, 139, 177, 179, 188, 190–96, 197, 198–204, 225, 232

*Xenopsylla cheopis* (Oriental rat flea), 40–41, 43, 73–74, 91, 92, 121, 144, 147, 158, 163, 186

Yeltsin, Boris, 220, 227
Ye-lu Ch'u-ts'ai, 101
Yersin, Alexandre, 180–83
*Yersinia enterocolitica*, 167–68
Yersinia murine toxin (YMT), 40–41
*Yersinia* outer proteins (Yops), 37, 47–48
*Yersinia pestis, see* plague
*Yersinia pseudotuberculosis*, 41, 42, 59, 167, 168
Yu Yueh, 179

Zhdanov, Victor M., 9–10, 18, 21–22, 25, 215
zoonosis, 3, 23, 171

# ABOUT THE AUTHOR

Wendy Orent is a leading freelance science journalist who writes for the *Los Angeles Times, The Washington Post, The New Republic, Discover,* and *Natural History*, among other publications. She earned a Ph.D. in anthropology from the University of Michigan in 1986, after which she spent six years as a research scientist in the anthropology department there. She is the coauthor of the English edition of *Biowarrior,* the memoirs of Igor V. Domaradskij, a principal designer of the Soviet bioweapons system.